Lecture Notes in Mathematics

A collection of informal reports and seminars
Edited by A. Dold, Heidelberg and B. Eckmann, Zürich

Series: California Institute of Technology, Pasadena
Adviser: C. R. DePrima

201

Jacobus H. van Lint
California Institute of Technology, Pasadena, CA/USA

Coding Theory

Second Printing

Springer-Verlag
Berlin · Heidelberg · New York 1973

AMS Subject Classifications (1970): 94 A 10

ISBN 3-540-06363-3 Springer-Verlag Berlin · Heidelberg · New York
ISBN 0-387-06363-3 Springer-Verlag New York · Heidelberg · Berlin

ISBN 3-540-05476-6 1. Auflage Springer-Verlag Berlin · Heidelberg · New York
ISBN 0-387-05476-6 1st edition Springer-Verlag New York · Heidelberg · Berlin

Offsetdruck: Julius Beltz, Hemsbach

PREFACE

These lecture notes are the contents of a two-term course given by me during the 1970-1971 academic year as Morgan Ward visiting professor at the California Institute of Technology. The students who took the course were mathematics seniors and graduate students. Therefore a thorough knowledge of algebra (a.o. linear algebra, theory of finite fields, characters of abelian groups) and also probability theory were assumed. After introducing coding theory and linear codes these notes concern topics mostly from algebraic coding theory. The practical side of the subject, e.g. circuitry, is not included. Some topics which one would like to include in a course for students of mathematics such as bounds on the information rate of codes and many connections between combinatorial mathematics and coding theory could not be treated due to lack of time. For an extension of the course into a third term these two topics would have been chosen.

Although the material for this course came from many sources there are three which contributed heavily and which were used as suggested reading material for the students. These are W. W. Peterson's Error-Correcting Codes ([15]), E. R. Berlekamp's Algebraic Coding Theory ([5]) and several of the AFCRL-reports by E. F. Assmus, H. F. Mattson and R. Turyn ([2], [3], [4] a.o.). For several fruitful discussions I would like to thank R. J. McEliece.

The extensive treatment of perfect codes is due to my own interest in this topic and recent developments. The reader who is familiar with coding theory will notice that in several places I have given a new treatment or new proofs of known theorems. Since coding theory is young there remain several parts which need polishing and several problems are still open. I sincerely hope that the course and these notes will contribute to the growing interest of mathematicians in this fascinating subject.

For her excellent typing of these lecture notes I thank Mrs. L. Decker.

Pasadena, March 1971. J. H. van Lint.

CONTENTS

CHAPTER IV: Important cyclic codes

CHAPTER V: Perfect codes

CHAPTER VI: Weight enumeration

NOTATION

P(a) and Prob (a) denote the probability of the event a.

Vectors are denoted by underlined symbols, e.g. $\underline{x}$, $\underline{y}$, $\underline{\theta}$.

$(\underline{a},\underline{b})$ is the usual inner product.

$\underline{a}\,\underline{b}$ is a product of vectors which is defined in (2.3.1).

$\mathcal{R}^{(n)}$ is the n-dimensional vector space (over a specified field GF(q)).

Systems with binary operations are denoted by giving the set and the operation, e.g. (GF(q)[x],+,) is the ring of polynomials with coefficients in GF(q) and addition denoted by + and multiplication denoted without a special symbol.

For matrices A and vectors $\underline{x}$ the transpose is A^T resp. $\underline{x}^T$.

If a and b are integers then $a|b$ means "a divides b".

If p is a prime then $p^e \| n$ means "$p^e | n$ and $p^{e+1} \nmid n$".

$A := B$ is used when the expression B defines A.

$A \subset B$ does not exclude $A = B$.

[ ] refers to the references at the end of the notes.

I. INTRODUCTION

1.1 Channels, noise, redundancy

The problems we shall be discussing, mostly of a mathematical nature, although sometimes belonging to the field of electrical engineering, are part of the theory of communication. In mentioning communication we think of many different things e.g. human speech, telephone conversations, storage devices like magnetic tape units for computers, high frequency radio, space communication links (e.g. telemetry systems in satellites), digital communication with computers etc. The problem concerns information coming from a source and going to a destination called the receiver through a medium we refer to as a channel (telephone, space). If the channel is "noiseless", i.e. every bit of information going in comes out unchanged, there is no problem but in practice this is not the case. Noise is added to the information and as a result errors are introduced by the channel. For example in telephone conversations there is cross-talk, thermal noise, impulsive switching noise, sometimes lightning causes noise; in radio we have static; magnetic tapes sometimes have the wrong digits after storage etc.

As an illustration we consider the following possible way of sending messages (think of teletype): we use 32 symbols namely a blank, the letters of our alphabet and a few punctuation marks (say 0 = blank, 1 = a, 2 = b, ...). Any message in English is converted into this code before entering the channel and back again afterwards as in the following model

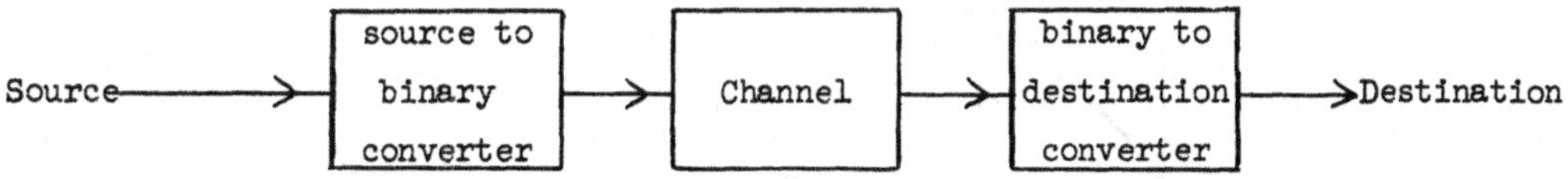

where e.g. the letter m is converted into 01101 and later back again. In the following we will no longer be concerned with the source and assume the information is presented as binary numbers or something analogous i.e. there is a finite set of information symbols at the source. For us the channel is an abstract thing in which noise is added according to specified rules. To better understand what actually happens we give one example.

Consider the two time functions given by the graphs below.

The first we identify with 0, the second with 1 and we let the functions denote the amplitude of a signal, e.g. the power emitted by a radio transmitter. Remark that the channel consists of several components, one being a binary to waveform converter. Now the letter m is transmitted through the channel as the following waveform:

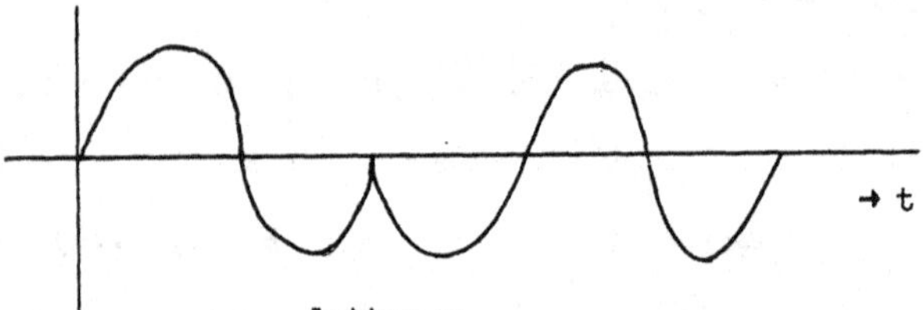

letter m

Due to the addition of noise the following wave is received.

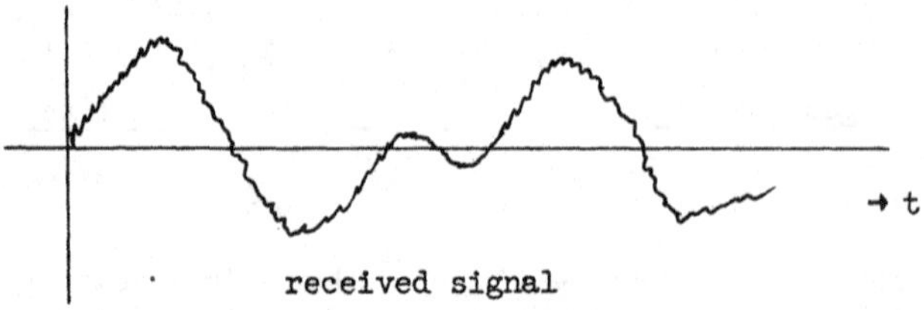

received signal

The receiver gets 01?01. The decision on the third bit is difficult. One possibility is to make no decision at all and to declare what is called an <u>erasure</u>. If the receiver must make a decision (<u>hard</u> decision) it is possible that an error is introduced, i.e. 01001 which is then converted into the letter i. In more refined situations information on how sure we are about each bit can be provided.

In the following we shall be concerned mostly with the case where errors are introduced. Even then there are several possibilities. The easiest to treat is the one where the errors are randomly distributed over the message. For many practical situations this is not a good model e.g. a flash of lightning will disturb several consecutive letters of a telephone conversation, errors on magnetic tape generally occur in bursts. In this case we speak of burst-errors. We shall treat these later. Normal spoken languages (in contrast to machine language, Morse code etc.) combat noise by using redundancy. Most words, especially long words, contain more letters than are necessary to recognize the word. This feature will be the basis of our theory resulting in the following communication scheme:

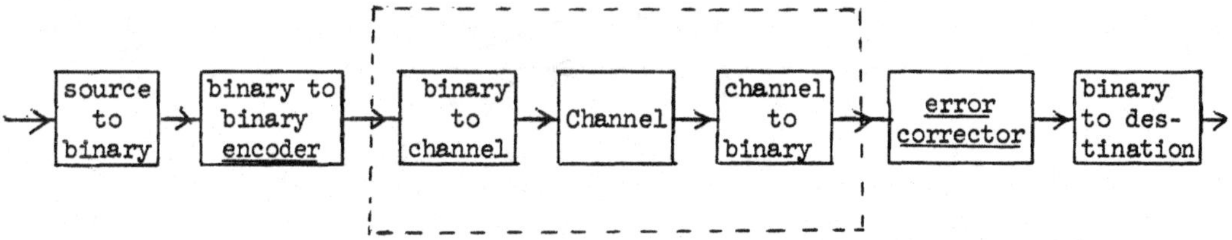

For many channels we could settle for error-detection instead of correction as in telephone where one can ask for a repetition of the message. We call this a two-way-channel or a channel with feedback. But if on a stored magnetic tape errors are detected it is too late to ask for a retransmission. Hence the main object of our study will be error-correction.

An example of an error-detecting code which is well known is the parity-check code used on paper tape for computers. There, after the information has been converted into binary digits (bits) one bit is added such that the mod 2 sum of the bits is 0 (e.g. 01101 is changed into 011011). The redundancy is one bit. This is called a parity-check. In case one error is introduced by the channel (no matter in which bit) this error is detected because the sum of the bits then has the wrong parity.

1.2 An introductory example

Consider a channel with an input alphabet $a_1, a_2, \ldots, a_k$ and an output alphabet $b_1, b_2, \ldots, b_\ell$. Suppose that each output letter depends statistically on the corresponding input letter only and according to a fixed probability. We write

$$P(b_j|a_i) := \text{probability that } b_j \text{ is received if } a_i \text{ is transmitted.}$$

This is called a discrete memoryless channel (DMC). We shall often consider the following special example of a DMC:

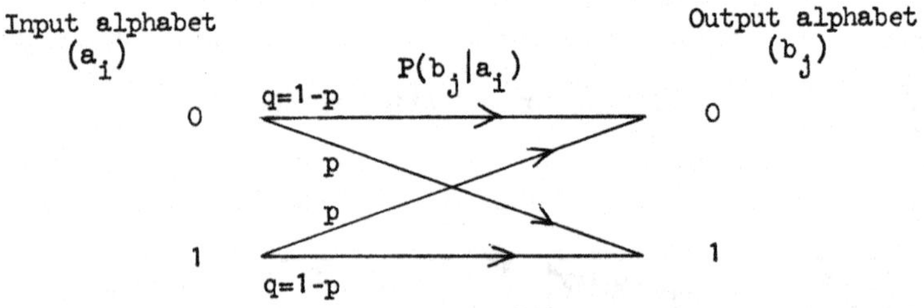

This is called the binary symmetric channel (BSC).

In our theory we shall assume that at the source each letter of the input alphabet has the same probability of occurring. Remark that if a certain symbol would occur more often than others it would be wise to use this knowledge in the construction of the code (as is done in shorthand).

Now suppose our BSC can handle 2A symbols per minute. We wish to use the channel to communicate to a receiver the result of an experiment carried out in the room we are in. A person is flipping a coin repeatedly at a speed of A times per minute. Every time "heads" comes up we can transmit a 0; if "tails" comes up a 1. To have some number to work with, assume the channel is poor and has $p = 0.02$. Then the receiver will get approximately 2% of the information incorrectly.

Of course there is time enough to repeat each message once. This is obviously useless as far as error-correction is concerned! Now consider the following code. We wait until the coin has been flipped twice and then transmit as follows:

Experiment result	Code message
HH	0000
TH	1001
HT	0111
TT	1110

Except for a delay at the beginning we'll keep up with the experiment. The receiver gets the following <u>decoding rule</u>: if anything different from these four messages is received then assume that one of the first three bits is in error. This rule uniquely assigns one code message to every possible sequence of four zeros and ones. Now let us do some computation:

Prob (4 bits received correctly) $= q^4$

Prob (one error among the first 3 bits) $= 3pq^3$

Hence Prob (correct decoding) $= q^4 + 3pq^3 \approx 0.9788$.

The receiver will get both experiments false if the last bit is received correctly and there are 2 or more errors among the first three bits. This has probability $p^3q + 3p^2q^2$. There remains probability p that one of the experiments will come through correctly. Therefore the receiver will get approximately 1.12% of the information in error. Quite an improvement! (But not best possible.)

Note that the receiver extracts two bits of information about the experiment from every four bits received. We say that the <u>information-rate</u> is $\frac{1}{2}$.

(1.2.1) DEFINITION: If we use a binary code of k words of length n (i.e. choose k out of a possible 2^n words) we shall say that the information-rate is $n^{-1} \log_2 k$.

We now continue with our example. We now wait until 3 flips of the coin have been executed and then transmit a sequence of 6 bits $a_1a_2\cdots a_6$ as follows. The outcome of the experiment is identical with $a_1a_2a_3$; $a_4 = a_2 + a_3$, $a_5 = a_3 + a_1$, $a_6 = a_1 + a_2$.

(All additions are mod 2). Just as in the first example the information rate is $\frac{1}{2}$. For the received word $(b_1,b_2,\ldots,b_6)$ we write $b_i = a_i + e_i$ and call $(e_1,e_2,\ldots,e_6)$ the <u>error-pattern.</u> Now we have

$$
\begin{aligned}
& b_2 + b_3 + b_4 = e_2 + e_3 + e_4 , \\
(1.2.2) \qquad & b_3 + b_1 + b_5 = e_3 + e_1 + e_5 , \\
& b_1 + b_2 + b_6 = e_1 + e_2 + e_6 .
\end{aligned}
$$

In (1.2.2) we compute the left-hand side from the received sequence. The rule is: of all possible error-patterns leading to the same left-hand side choose the one with the least number of errors. This method is called <u>maximum-likelihood decoding</u>. The effect is as follows:

a) $\begin{pmatrix}0\\0\\0\end{pmatrix}$; no errors , probability is q^6,

b) $\begin{pmatrix}0\\0\\1\end{pmatrix},\begin{pmatrix}0\\1\\0\end{pmatrix},\begin{pmatrix}1\\0\\0\end{pmatrix}$; one error among a_4,a_5,a_6, probability is $3q^5p$, error is corrected,

c) $\begin{pmatrix}1\\1\\0\end{pmatrix},\begin{pmatrix}0\\1\\1\end{pmatrix},\begin{pmatrix}1\\0\\1\end{pmatrix}$; one error among a_1,a_2,a_3, probability is $3q^5p$, error is corrected.

d) $\begin{pmatrix}1\\1\\1\end{pmatrix}$; assume e_1,e_4 are false , probability is q^4p^2.

Apparently any single error leads to correct decoding! What about double errors? We must distinguish between

α) 2 errors among first three, say $e_2 = e_3 = 1$. This leads to $\begin{pmatrix}0\\1\\1\end{pmatrix}$, decoded as if the error pattern were (1,0,0,0,0,0). Final result: all three wrong!

β) 2 errors among last three, say $e_4 = e_5 = 1$. This leads to $\begin{pmatrix}1\\1\\0\end{pmatrix}$, decoded as if the error pattern were (0,0,1,0,0,0). Final result: two correct, one false.

γ) one error in first three, one in last three. This leads to either $\begin{pmatrix}1\\1\\1\end{pmatrix}$ or $\begin{pmatrix}0\\0\\1\end{pmatrix}$ which are decoded as if the error pattern were (1,0,0,1,0,0) resp. (0,0,0,0,0,1). In the last case two experiment-results are interpreted correctly. In the first case either both errors are corrected as mentioned in d) above or two more are introduced. Since p is small we do not calculate what happens if 3 errors occur.

The average percentage of incorrect information at the receiver is now about 0.29%. The normal question to ask at this point is: "Can we get the percentage of error down as low as we wish using information rate 1/2?" We shall now answer this question.

First a definition:

(1.2.3) DEFINITION: Consider the space $\{0,1\}^n$ consisting of all sequences of n zeros and ones. In this n-dimensional vector space (addition mod 2) we define the Hamming-distance $d_H(\underline{x},\underline{y})$ of two vectors $\underline{x}$ and $\underline{y}$ to be the number of places in which these vectors differ.

Note that with d_H as distance $\{0,1\}^n$ is a metric space.

(1.2.4) DEFINITION: For $\rho > 0$ and $\underline{x} \in \{0,1\}^n$ we define the sphere of radius ρ with center at $\underline{x}$ by

$$S_\rho(\underline{x}) := \{\underline{y} \in \{0,1\}^n \mid d_H(\underline{x},\underline{y}) \leq \rho\}.$$

If, on continuing with our example, we wait until m flips of the coin have been executed and then transmit 2m bits, suppose we succeed in choosing the code in such a way that for each pair of distinct code words $\underline{x}$ and $\underline{y}$ we have $d_H(\underline{x},\underline{y}) \geq 5$. What does this mean? The effect is that if a received word contains less than 3 errors the transmitted word can be determined by maximum likelihood decoding because there is exactly one code word at a distance $d_H \leq 2$ from the received word (by the triangle inequality). The probability of correct decoding is now $q^{2m} + \binom{2m}{1}q^{2m-1}p + \binom{2m}{2}q^{2m-2}p^2 \approx 1 - \binom{2m}{3}p^3$ if m is not too large and p is small. So, much depends on how large m must be to achieve this proerty of correcting 2 errors. We shall not attempt to find m but we shall find the answer to our question in section 1.4.

1.3 Some definitions in information theory

Let X denote a probability space consisting of the set $\{a_1,a_2,\ldots,a_k\}$ with probabilities $P_X(a_i)$, $i = 1,2,\ldots,k$ and let Y be the probability space consisting of $\{b_1,b_2,\ldots,b_\ell\}$ with probabilities $P_Y(b_j)$, $j = 1,2,\ldots,\ell$. Also let a probability measure P_{XY} be given on the product space of pairs (a_i,b_j). Note that there are several relations connecting these probabilities e.g.

(1.3.1)
$$P_X(a_i) = \sum_{j=1}^{\ell} P_{XY}(a_i,b_j) \quad (i = 1,\ldots,k),$$

if we interpret (a_i,b_j) as the outcomes of a 2-outcome experiment, X denoting the first outcome, etc. As usual we denote the probability of the occurrence of the event a_i, given that the outcome of the second alternative is b_j by $P(a_i|b_j)$ or $P_{X|Y}(a_i|b_j)$. Clearly

(1.3.2)
$$P_{X|Y}(a_i|b_j) = \frac{P_{XY}(a_i,b_j)}{P_Y(b_j)} .$$

We recall that the events $x = a_i$ and $y = b_j$ are called statistically independent if

(1.3.3)
$$P_{XY}(a_i,b_j) = P_X(a_i)P_Y(b_j),$$

i.e.
$$P_{X|Y}(a_i|b_j) = P_X(a_i).$$

Now in the situations we are interested in the a_i will be inputs of a channel and the b_j outputs. Usually we shall consider situations where all the $P_X(a_i)$ are equal but right now we do not assume that yet. Of course in this case independence is the last thing we want. In fact if the output is some b_j we would like to be certain about the input (noiseless channel). To measure what a certain output tells us we introduce the quantity

(1.3.4) DEFINITION:
$$I_{X,Y}(a_i,b_j) := \log_2 \frac{P_{X|Y}(a_i|b_j)}{P_X(a_i)} = \log_2 \frac{P_{XY}(a_i,b_j)}{P_X(a_i)P_Y(b_j)} ,$$

and call this the information provided about the event $x = a_i$ by the occurrence of the event $y = b_j$ or also the mutual information of $x = a_i$ and $y = b_j$.

Examples: (a) In the case of independence no information is provided and $I = 0$; (b) For a noiseless channel the probability spaces X and Y are identical and $I_{X,Y}(\xi,\xi) = \log_2 \frac{1}{P_X(\xi)}$.

(1.3.5) DEFINITION: $I_X(a_i) := \log_2 \frac{1}{P_X(a_i)}$ is called the <u>self-information</u> of the event $x = a_i$.

Example: If X is the set of all binary sequences of length n, each having probability 2^{-n} then the self-information of one sequence is n, i.e. the number of bits of that sequence. For this reason information is generally measured in bits.

(1.3.6) DEFINITION: The quantity

$$I(X;Y) := \sum_{i=1}^{k} \sum_{j=1}^{\ell} P_{XY}(a_i,b_j) I_{X,Y}(a_i,b_j)$$

is called the <u>average mutual information</u>.

Analogously

(1.3.7) DEFINITION: The quantity

$$H(X) := \sum_{i=1}^{k} P_X(a_i) \log \frac{1}{P_X(a_i)} = -\sum_{i=1}^{k} P_X(a_i) \log P_X(a_i)$$

is called the <u>average self-information</u> or <u>entropy</u> of the probability space X.

We shall not go into the connection with the use of the word entropy in thermodynamics. A. I. Khinchin has shown (cf. [8]) that if one wishes to measure uncertainty and requires this measure to have a few natural properties there is only one function with these properties, namely the entropy function. Now as was suggested above interpret X and Y as input and output of a discrete memoryless channel. Here the $P_{Y|X}(b_j|a_i)$ depend on the channel. We can still choose the probabilities $P_X(a_i)$ $(i = 1,2,\dots,k)$ any way we like. This leads us to:

(1.3.8) DEFINITION: The maximum average mutual information $I(X;Y)$ over all possible input probabilities is called the <u>capacity</u> C of the DMC.

It should now be plausible that C is a measure for the amount of information the channel can handle in one use.

Let us compute C for the BSC with error probability p. Let $P_X(0) = x$ $(0 \leq x \leq 1)$. Then by (1.3.4) and (1.3.6) we have

$$I(X;Y) = x\left\{(1-p)\log \frac{1-p}{x(1-p)+(1-x)p} + p \log \frac{p}{xp+(1-x)(1-p)}\right\}$$

$$+ (1-x)\left\{(1-p)\log \frac{1-p}{xp+(1-x)(1-p)} + p \log \frac{p}{x(1-p)+(1-x)p}\right\}$$

$$= p \log p + q \log q - \{x(1-p) + (1-x)p\}\log\{x(1-p) + (1-x)p\}$$

$$- \{xp + (1-x)(1-p)\}\log\{xp + (1-x)(1-p)\}$$

and this is maximal if $x = \frac{1}{2}$ in which case we find

(1.3.9) The capacity of the BSC with error probability p is $1 + p \log p + q \log q$ (where $q := 1-p$ as usual).

Sometimes (1.3.9) is given as definition with no further comment. We hope the above explanation makes the word capacity a little less vague.

1.4 Shannon's Fundamental Theorem

Consider once again a BSC with error probability p, $(0 < p < \frac{1}{2})$. Suppose we have a code consisting of M vectors chosen from $\{0,1\}^n$ with some decoding rule. Let P_i denote the probability that an error occurs (after decoding) if $\underline{x}_i$ is transmitted. Then the probability of error when using this code is

(1.4.1) $$P_{error} = M^{-1} \sum_{i=1}^{M} P_i.$$

We now define

(1.4.2) $P^*(M,n,p)$:= minimum of P_{error} over all codes with the given parameters.

We can now formulate Shannon's first theorem which is the basis of the rest of this course. Let C be the capacity of the BSC, defined in (1.3.9).

(1.4.3) THEOREM (Shannon): Let $M_n := 2^{[Rn]}$ where $0 < R < C$. Then $P^*(M_n,n,p) \to 0$ if $n \to \infty$.

Note that this means that there is a sequence of codes with information rate tending to R and error probability tending to 0. Or in other words: given $\epsilon > 0$ and $R < C$ there is a code with rate $> R$ and error-probability $< \epsilon$.

We now prove this theorem. Before giving the main idea of the proof we isolate a number of the more technical details to be used in the main part of the proof.

If a code word is transmitted over the channel then the probability of a specific error pattern (at the receiver) with w errors is $p^w q^{n-w}$, i.e. this probability depends only on the number of errors. Remark that $P(\underline{x}|\underline{y}) = P(\underline{y}|\underline{x})$. The number of errors in received words is a random variable with expected value np and variance $np(1-p)$. If $b := \left(\frac{np(1-p)}{\epsilon/2}\right)^{1/2}$ then by the Bienaymé-Chebyshev inequality

(1.4.4) $$\text{Prob } (w > np + b) \le \frac{1}{2}\epsilon.$$

Let $\rho := [np + b]$. By (1.2.4) the number of points in a sphere $S_\rho(\underline{x})$ is

(1.4.5) $$|S_\rho(\underline{x})| = \sum_{w \le \rho} \binom{n}{w} < \frac{1}{2} n \binom{n}{\rho} \le \frac{1}{2} n \cdot \frac{n^n}{\rho^\rho (n-\rho)^{n-\rho}} .$$

The last inequality follows from $n^n = (\rho + n-\rho)^n = \cdots + \binom{n}{\rho}\rho^\rho(n-\rho)^{n-\rho} + \cdots$.

We have

(1.4.6) $$\frac{\rho}{n} \log \frac{\rho}{n} = \frac{[np + b]}{n} \log \frac{[np + b]}{n} = (p + \frac{b}{n}) \{\log p + \log (1 + \frac{b}{np})\} + O(\frac{1}{n}) =$$
$$= p \log p + O(n^{-\frac{1}{2}}).$$

and analogously $(1 - \frac{\rho}{n}) \log (1 - \frac{\rho}{n}) = q \log q + \mathcal{O}(n^{-1/2})$.

We introduce two auxiliary functions. If $\underline{u} \in \{0,1\}^n$, $\underline{v} \in \{0,1\}^n$ we define:

(1.4.7) $$f(\underline{u},\underline{v}) := \begin{cases} 0 & \text{if } d_H(\underline{u},\underline{v}) > \rho , \\ 1 & \text{if } d_H(\underline{u},\underline{v}) \le \rho . \end{cases}$$

If $\underline{x}_i$ is any vector in the code and $\underline{y} \in \{0,1\}^n$ then

(1.4.8) $$g_i(\underline{y}) := 1 - f(\underline{y},\underline{x}_i) + \sum_{j \ne i} f(\underline{y},\underline{x}_j).$$

Remark that $g_i(\underline{y}) = 0$ if $\underline{x}_i \in S_\rho(\underline{y})$ and no other code word is in this sphere and otherwise $g_i(\underline{y}) \geq 1$.

Now we come to the main part of our proof. We choose the code words $\underline{x}_1$, $\underline{x}_2$, ..., $\underline{x}_M$ at random from $\{0,1\}^n$ and use these to transmit M possible messages. The decoding rule for the receiver is as follows. If $\underline{y}$ is received and there is one code word $\underline{x}_i$ with distance $\leq \rho$ to $\underline{y}$ then $\underline{y}$ is decoded into $\underline{x}_i$. Otherwise an error is declared (or one could always take $\underline{x}_1$ in this case). Again, let P_i denote the probability of incorrect decoding when $\underline{x}_i$ is transmitted. We have, by the remark following (1.4.8),

$$P_i \leqq \sum_{\underline{y} \in \{0,1\}^n} P(\underline{y}|\underline{x}_i) g_i(\underline{y}) =$$

$$= \sum_{\underline{y}} P(\underline{y}|\underline{x}_i)\{1 - f(\underline{y},\underline{x}_i)\} + \sum_{\underline{y}} \sum_{j \neq i} f(\underline{y},\underline{x}_j) P(\underline{y}|\underline{x}_i).$$

The first sum on the right-hand side is just the probability that $\underline{y}$ is not in $S_\rho(\underline{x}_i)$. This number does not depend on $\underline{x}_i$ but only on ρ. Call it α_ρ. By (1.4.4) we have $\alpha_\rho \leq \frac{1}{2}\epsilon$.

Now by (1.4.1)

$$P_{error} \leq \frac{1}{2}\epsilon + M^{-1} \sum_{i=1}^{M} \sum_{\underline{y}} \sum_{j \neq i} P(\underline{y}|\underline{x}_i) f(\underline{y},\underline{x}_j).$$

We now calculate the expected value of the right-hand side and remark that $P^*(M,n,p)$ can not be larger. Since the $\underline{x}_i$ are chosen at random they are independent variables. Hence we have

$$P^*(M,n,p) \leq \frac{1}{2}\epsilon + M^{-1} \sum_{i=1}^{M} \sum_{\underline{y}} \sum_{j \neq i} E(P(\underline{y}|\underline{x}_i)) E(f(\underline{y},\underline{x}_j)) =$$

$$= \frac{1}{2}\epsilon + M^{-1} \sum_{i=1}^{M} \sum_{\underline{y}} \sum_{j \neq i} E(P(\underline{y}|\underline{x}_i)) \frac{|S_\rho|}{2^n} =$$

$$= \frac{1}{2}\epsilon + (M-1)2^{-n}|S_\rho|.$$

Now take logarithms (base 2), use (1.4.5) and (1.4.6) and divide by n. We find

(1.4.9) $$n^{-1} \log \{P^*(M,n,p) - \tfrac{1}{2}\varepsilon\} \leq n^{-1} \log M - C + O(n^{-\frac{1}{2}})$$

By definition of M_n we have $n^{-1}\log M_n - C + \mathcal{O}(n^{-1/2}) < -\beta < 0$ for $n > n_0$, i.e.

$$P^*(M_n,n,p) < \tfrac{1}{2}\varepsilon + 2^{-\beta n} \text{ for } n > n_0 .$$

This completes the proof.

1.5 Problems

(1.5.1) Consider the vectorspace $\{0,1\}^6$ with Hamming distance. What is the number of points in a sphere of radius 1? Is it possible to find 9 vectors such that for each pair $\underline{x},\underline{y}$ ($\underline{x} \neq \underline{y}$) $d_H(\underline{x},\underline{y}) \geq 3$?

(1.5.2) Consider a BSC with error probability p ($0 < p < 1/2$). We wish to transmit two possible messages 0 and 1 by using a repetition code of length $2n + 1$, i.e. $0 = (0,0,\ldots,0)$ and $1 = (1,1,1,\ldots,1)$. Let P_n denote the probability of incorrect decoding. Prove $\lim_{n \to \infty} P_n = 0$.

(1.5.3) Consider a BSC with 10% error probability. Four possible messages are coded using words from $\{0,1\}^6$. What error probability can be achieved?

(1.5.4) We wish to transmit 0's and 1's over a binary erasure channel with 10% erasure probability (i.e. if a 0 or 1 is transmitted there is probability 0.9 that it is received correctly and probability 0.1 that a ? is received). Transmission with information rate 1/2 is acceptable. Does repeating each bit help? Can we do any better?

(1.5.5) Consider the code with 8 code words $\in \{0,1\}^6$ defined by $a_4 = a_2 + a_3$, $a_5 = a_1 + a_3$, $a_6 = a_1 + a_2$. Let this code be used over an erasure channel with probability p of erasure. Determine the probability of correct decoding. Compare this with repetition of each bit.

(1.5.6) The entropy of a discrete probability space is maximal if all probabilities are equal. Prove this.

(1.5.7) Consider the alphabet $\{0,1,2\}$. Try to find 9 four letter words such that they form a single-error-correcting code.

II. LINEAR CODES

2.1 General Theory

In this chapter we construct codes for a discrete memoryless channel for which the input and output alphabets are a set of q symbols where q is a power of a prime, $q = p^{\alpha}$. In this case we can identify these alphabets with the elements of GF(q). We are going to construct block codes, i.e. codes in which all code words have the same length n which is called the block length or word length.

Let $\mathcal{R}$ be the vector space of dimension n over GF(q). A subset V of $\mathcal{R}$ is an e-error-correcting-code if $\forall_{\underline{x} \in V} \forall_{\underline{y} \in V}[(\underline{x} \neq \underline{y}) \Rightarrow (d_H(\underline{x},\underline{y}) \geq 2e+1)]$ where the Hamming distance d_H is defined as in (1.2.3). (If certain errors are more likely than others, Hamming distance is not a good distance function to use!)

(2.1.1) DEFINITION: A k-dimensional linear subspace V of $\mathcal{R}$ is called a linear code or (n,k)-code over GF(q).

Linear codes are group codes (V is a subgroup of $(\mathcal{R},+)$). In general a group-code V is a subgroup of the direct product of n copies of an additive abelian group G. Since most of the codes we shall be studying in this course are linear codes it will be useful to go into the general theory first and find simple representations of such codes.

(2.1.2) DEFINITION: The Hamming-weight $w(\underline{x})$ of a vector $\underline{x} \in \mathcal{R}$ is the number of coordinates different from 0.

The first advantage of linear codes compared to other codes is the following property.

(2.1.3) For a linear code $V \subset \mathcal{R}$ the minimum distance of different code vectors is equal to the minimum weight of all non-zero code vectors.

Since the minimum distance of the code determines the number of errors that can be corrected when using maximum likelihood decoding, it is convenient to have this easy way of finding the minimum distance. Of course if e.g. q = 2, n = 100, k = 80 one cannot inspect all code words to find the minimum weight!

In the following we shall generally take q = 2 as an example but the theory holds for all fields GF(q). One advantage of GF(2) is that we can change any - sign into + if we prefer that.

We shall now look at two ways of describing the code V. The first is to take a basis of the linear subspace V and to form the matrix G where the rows of G are the basis vectors of V. The matrix G is called the generator matrix of the code. E.g.

$$G := \begin{pmatrix} 1 & 0 & 0 & 1 & 1 \\ 0 & 1 & 0 & 1 & 0 \\ 0 & 0 & 1 & 0 & 1 \end{pmatrix}$$

is the generator matrix of a (5,3)-code over GF(2).

Since the property we are most interested in is possible error-correction and this property is not changed if in all code words we interchange two symbols (e.g. the first and second letter of each code word) we shall call two codes equivalent if one can be obtained by applying a fixed permutation to the words of the other. With this in mind we see that for every linear code there is an equivalent code which has a generator matrix of the form

$$G = (I_k P)$$

where I_k is the k by k identity matrix and P is a k by (n-k) matrix. The example given above is of this type. This is often called the reduced echelon form of G.

At this point we should remark that as far as error-correction is concerned the code does not change either if in any coordinate place we change all o's into 1's and vice-versa. The resulting code is no longer linear. In fact what this transformation amounts to in the binary case is adding a fixed vector to all code words i.e. looking at a coset of the linear subspace instead of the subspace itself. (This more general equivalence will be used in Section 5.1).

Now a well known theorem from linear algebra states that if V is a k-dimensional linear subspace of $\mathcal{R}$ then there is an (n-k)-dimensional subspace $V^{\perp}$ of $\mathcal{R}$ such that $\forall_{\underline{x} \in V} \forall_{\underline{y} \in V^{\perp}} [(\underline{x},\underline{y}) = 0]$ where $(\underline{x},\underline{y})$ denotes the usual inner product. $V^{\perp}$ is called the orthogonal complement of V. We warn the reader that it is no longer true, as in spaces over $\mathbb{R}$, that each vector in $\mathcal{R}$ is the sum of a vector from V and a vector from

$V^{\perp}$. In fact it is possible that V and $V^{\perp}$ coincide. The code $V^{\perp}$ is called the dual of V. If $V = V^{\perp}$ then V is called self-dual. To avoid confusion we warn the reader that often, when speaking of the dual code the order of the symbols is reversed. Why this is done in certain instances will become clear later.

A generator matrix H of the dual V is called a parity-check matrix of the code V. For instance for the example given above

$$H := \begin{pmatrix} 1 & 1 & 0 & 1 & 0 \\ 1 & 0 & 1 & 0 & 1 \end{pmatrix}$$

is a parity-check matrix. If $G = (I_k P)$ then we can take $H = (-P^T I_{n-k})$. Remember that the single parity-check code mentioned in Section 1.1 had the property that $(a_1, a_2, \ldots, a_6)$ was a code word iff $a_1 + a_2 + \cdots + a_6 = 0$ (addition in GF(2)). This has now been generalized. A word $\underline{x} \in \mathcal{R}$ is in the code V iff it is orthogonal to every row of H, i.e.

$$(2.1.4) \qquad (\underline{x}\, H^T = \underline{0}) \Leftrightarrow (\underline{x} \in V).$$

Here (2.4) is a system of n - k linear equations which must be satisfied if $\underline{x}$ is a code word. These are called the parity-check equations. For any $\underline{x} \in \mathcal{R}$ we call $\underline{x}\, H^T$ the syndrome of $\underline{x}$. Then $\underline{x} \in V$ iff the syndrome is $\underline{0}$. Using the reduced echelon form makes it very clear what is actually happening. We can make a code word $(a_1, a_2, \ldots, a_n)$ by choosing $a_1, a_2, \ldots, a_k$ arbitrarily. Then either G or H tells us the rest of the word namely (using G) we have to take a_1 times the first row of $G + a_2$ times the second row etc., respectively (using H) we have $a_{k+1} = -p_{11}a_1 - p_{21}a_2 \cdots -p_{k1}a_k$ etc. according to (2.4). So the bits $a_1, a_2, \ldots, a_k$ carry the information and the remaining n - k bits are the redundant part which is there to help correct errors. The rate of information (q = 2) is apparently k/n. We have here an example of what is called a systematic code i.e. one for which a subset of symbols can take on arbitrary values (these are the information symbols) and the other symbols are then determined (parity-check symbols). The example described by (1.2.2) is a (6,3) code over GF(2) with parity check matrix

$$H = \begin{pmatrix} 0 & 1 & 1 & 1 & 0 & 0 \\ 1 & 0 & 1 & 0 & 1 & 0 \\ 1 & 1 & 0 & 0 & 0 & 1 \end{pmatrix} \quad .$$

The reader who now thinks we have embarked on a study of a trivial part of linear algebra can try to find the minimum weight of a linear code from the generator G. This should be sufficient to change his mind!

Now consider the problem of correcting errors in a linear code. Let us take a particular error-pattern, i.e. a vector in $\mathcal{R}$ and add it to all the code words. The resulting set is a coset of V in the additive group $(\mathcal{R},+)$. If $\underline{e}$ is the error-pattern the coset is $\underline{e} + V := \{\underline{v} + \underline{e} \mid \underline{v} \in V\}$. If $\underline{v}_1$ and $\underline{v}_2$ are in the code then $(\underline{v}_1 + \underline{e})H^T = (\underline{v}_2 + \underline{e})H^T = \underline{e}H^T$, i.e. two vectors in the same coset have the same syndrome. On the other hand, if $\underline{x}H^T = \underline{y}H^T$ then $(\underline{x} - \underline{y})H^T = \underline{0}$, i.e. $\underline{x} - \underline{y} \in V$ and hence $\underline{x}$ and $\underline{y}$ are in the same coset of V. We have proved:

(2.1.5) THEOREM: Two vectors $\underline{x}$ and $\underline{y}$ are in the same coset of V iff they have the same syndrome.

The next important remark is that the error-pattern $\underline{e}$ is itself in the coset $\underline{e} + V$ and any vector in this coset could be the error pattern. To prove the last statement take $\underline{v}_1 \in V$. Then $(\underline{v}_1 + \underline{e}) + V = \underline{e} + V$ since $\underline{v}_1 + V = V$. The last remark now tells us how to proceed when using maximum likelihood decoding. If we wish to find an error-pattern which changes the words of V into the words of a particular coset then the candidates are the words of minimum weight in this coset. For each coset we choose such a word and call this word the coset leader. Consider the following array: (i) in the first row we write the words of the code, starting with $\underline{0}$ (which is the leader of the coset V); (ii) in the first column we write the coset leaders $\underline{0}, \underline{e}_1, \underline{e}_2, \ldots$; (iii) in the row with leader $\underline{e}_i$ we write $\underline{e}_i + \underline{v}_j$ under the code word $\underline{v}_j$. This is called the standard array. As an example we take the first code used in Section 1.2. This is a linear code with $G = \begin{pmatrix} 1 & 0 & 0 & 1 \\ 0 & 1 & 1 & 1 \end{pmatrix}$. The standard array corresponding to the decoding rule we used is:

0 0 0 0	1 0 0 1	0 1 1 1	1 1 1 0
1 0 0 0	0 0 0 1	1 1 1 1	0 1 1 0
0 1 0 0	1 1 0 1	0 0 1 1	1 0 1 0
0 0 1 0	1 0 1 1	0 1 0 1	1 1 0 0 .

For the receiver this array is a dictionary to be used as follows: decode a received word into the code word at the top of the column containing the received word. Note that in this example the coset leaders are uniquely determined except in the second row where we could have taken 0001.

For any code a complete dictionary is the least sophisticated way of providing a decoding rule. We can now find something much simpler for linear codes by recalling that words in the same row of the standard array all have the same syndrome, this being the syndrome of the error pattern which is the coset leader. Hence

(2.1.6) THEOREM. For a linear code a maximum likelihood decoding rule is completely described if a list of coset leaders with their syndromes is given.

The way to use (2.1.6) is: (i) If $\underline{x}$ is received compute the syndrome $\underline{x}H^T$; (ii) look up the syndrome in the table to find the error pattern $\underline{e}$; (iii) $\underline{x} - \underline{e}$ is the code word.

For an (n,k)-code over GF(2) a complete dictionary consists of all possible 2^n words whereas the list of (2.1.6) has only 2^{n-k} words and their syndromes. In practice good codes are long e.g. $n = 100$, $k = 80$ and then the list of (2.1.6) would still be much too long to be of practical use.

We have seen that the coset leader is not always uniquely defined. To change this we use an idea of E. Prange and D. Slepian. Number the elements of GF(q) from 1 to q giving the field element 0 the number q. Now in every coset choose as the leader the minimal weight element which comes first lexicographically. We can now prove an interesting theorem. First we define

(2.1.7) DEFINITION: If $\underline{x} \in \mathcal{R}$, $\underline{y} \in \mathcal{R}$ then we define $\underline{x} \prec \underline{y}$ by

$$\underline{x} \prec \underline{y} :\Leftrightarrow \forall_{i, 1 \le i \le n} [x_i = y_i \text{ or } x_i = 0].$$

Then we have

(2.1.8) THEOREM: If $\underline{y}$ is a coset leader and $\underline{x} \prec \underline{y}$ then $\underline{x}$ is a coset leader.

Proof: It is sufficient to prove the theorem for the case $w(\underline{y} - \underline{x}) = 1$. In this case let $\underline{y} - \underline{x} = \underline{e}$. Every vector in the coset $\underline{y} + V$ can be obtained by adding $\underline{e}$ to a suitable vector in $\underline{x} + V$. Since all vectors in $\underline{y} + V$ have weight $\geq w(\underline{y})$ no vector in $\underline{x} + V$ can have weight less than $w(\underline{y}) - 1 = w(\underline{x})$, i.e. $\underline{x}$ has minimal weight in its coset. Now let $e_k \neq 0$ and $e_i = 0$ for $i \neq k$. If $\underline{a}$ is a minimal-weight vector in $\underline{x} + V$ then $\underline{a} + \underline{e}$ is a minimal weight vector in $\underline{y} + V$ and hence $a_k = 0$. If $\underline{a} \neq \underline{x}$ then $\underline{x} + \underline{e}$ is lexicographically after $\underline{y}$ and hence $\underline{a}$ is lexiocraphically after $\underline{x}$. This proves the theorem.

A decoding algorithm called step-by-step decoding is based on the above definition of standard array and (2.1.8). We refer the interested reader to [15], Chapter 3.

2.2 Hamming codes

Let H be the parity-check matrix of a binary linear code. If a code word $\underline{x}$ is transmitted and received with error-pattern $\underline{e}$ (i.e. $\underline{x} + \underline{e}$ is received) then the syndrome as defined in the previous section is $\underline{e}H^T$. This syndrome is just the (transpose of the) sum of those columns of H corresponding to places where errors have occurred. Therefore an error in a place corresponding to a column of zeros in H does not influence the syndrome and hence such an error is not even detected! If H has two identical columns and in the corresponding places errors are made the syndrome is again $\underline{0}$, i.e. these two errors are not detected. If, on the other hand, all columns of H are different then a single error in the k-th coordinate results in the syndrome being the k-th column of H. Therefore the single error can be localized and corrected. Therefore every weight one vector is a coset leader. An other way of seeing this is to remark that a linear combination of columns of H can be $\underline{0}$ only if the number of terms with nonzero coefficient is at least three or in other words $\underline{x}H^T = \underline{0}$ implies $w(\underline{x}) \geq 3$. Therefore the minimum distance of the code is at least 3 and the code is single-error-correcting. We have thus given two proofs of

(2.2.1) <u>THEOREM</u>: <u>A linear binary code can correct all patterns of not more than one error iff all columns of the parity-check matrix are different and nonzero.</u>

If H has r rows, i.e. the number of parity check symbols is $\leq r$ then clearly H can have at most $2^r - 1$ columns if they are all to be different and nonzero.

(2.2.2) DEFINITION: A binary linear code for which the columns of the parity check matrix H are the binary representations of 1, 2, ..., 2^r-1 is called the $(2^r-1, 2^r-1-r)$ Hamming code.

The (7,4) Hamming code has the parity check matrix

$$H := \begin{pmatrix} 0 & 0 & 0 & 1 & 1 & 1 & 1 \\ 0 & 1 & 1 & 0 & 0 & 1 & 1 \\ 1 & 0 & 1 & 0 & 1 & 0 & 1 \end{pmatrix} .$$

If a word $\underline{x}$ is received and $\underline{x}H^T$ is the binary representation of the integer k then the most likely error pattern is one error in the k-th place. Note that this is a tremendous improvement compared to (2.1.6) since now knowledge of H is sufficient for the decoding rule!

If in the case of the (7,4) code we wish to generate the code words we apply the permutation (4,5)(6,2)(1,7) to H to obtain the form with I_3 at the end, obtain G as described in Section 2.1 and permute back. We find

$$G = \begin{pmatrix} 1&1&0&1&0&0&1 \\ 0&1&0&1&0&1&0 \\ 1&1&1&0&0&0&0 \\ 1&0&0&1&1&0&0 \end{pmatrix} \text{ or after some linear combination } \begin{pmatrix} 1&0&0&0&0&1&1 \\ 0&1&0&0&1&0&1 \\ 0&0&1&0&1&1&0 \\ 0&0&0&1&1&1&1 \end{pmatrix} .$$

From this last form of G it is obvious that all code words have weight ≥ 3. The Hamming codes have a remarkable property. Not only are the spheres with radius one and centers at the code words disjoint but they also fill the whole space $\mathcal{R}$. Such a code is called <u>perfect</u>.

(2.2.3) DEFINITION: If an e-error-correcting code $V \subset \mathcal{R}$ has the property

$$\bigcup_{\underline{x} \in V} S_e(\underline{x}) = \mathcal{R}$$

then the code is called <u>perfect</u>.

If r is the number of parity-check symbols of a Hamming code then the block-length n is $2^r - 1$. The number of information symbols is $2^r - 1 - r$ and the rate is therefore

$1 - \frac{r}{2^r - 1}$ which is nearly 1 for large r. The proof that a Hamming code is perfect follows from the equality

$$2^n = (1+n)\cdot 2^k \quad \text{if} \quad n = 2^r - 1 \text{ and } k = 2^r - 1 - r.$$

Of course a consequence of the fact that a Hamming code is perfect is that it can not detect any pattern of two errors! A very simple device will remedy this, namely the addition (to every word) of one more check symbol which is the sum of the other symbols. If we call this symbol a_0 then the new parity check matrix is

$$H^* := \begin{pmatrix} 1 & 1\ 1\ 1\ 1 \cdots 1 \\ 0 & \\ 0 & \\ 0 & H \\ \vdots & \\ 0 & \end{pmatrix}.$$

Since this _extended_ code obviously has minimum weight 4 it can detect all patterns of 2 errors. One way of using such a code is to employ an incomplete decoding rule which corrects all error patterns of weight ≤ 1 and which declares a _decoding failure_ in case 2 errors are detected.

The idea behind Hamming codes is not restricted to binary codes. An as example we take GF(3) and construct all nonzero columns of r elements of GF(3) with the restriction now that (starting from the top) the first nonzero element is always 1. The number of columns of this type is $n := \frac{1}{2}(3^r - 1)$. Once again a linear combination of two of these columns, with coefficients 1 or 2 cannot be $\underline{0}$. Therefore if we take H to be the matrix with these columns then $\underline{x}H^T = \underline{0}$ implies $w(\underline{x}) \geq 3$, i.e. H is the parity check matrix of a single error correcting code over GF(3). Once again $|S_1| = 1 + 2n = 3^r = 3^{n-k}$ where k is the dimension of the code. Therefore the code is perfect. The same proof holds with q instead of 3.

(2.2.4) _THEOREM_: _Hamming codes over a field_ GF(q) _are perfect codes._

The (13,10) ternary Hamming code provides a solution to the _football-pool problem_ for 13 matches. (In some countries this is the actual number of matches in the pool!) In a football pool one forecasts the outcome of a number of (Soccer)-matches:

0 = draw, 1 = win, 2 = lose (for home team). To be sure of first prize (all correct) there is no alternative but to enter all 3^{13} possible forecasts. A priori it is clear that for second prize (12 correct) it suffices to enter 3^{12} forecasts but in fact 3^{10} are enough! Since the (13,10) Hamming code is perfect every word in the 13-dimensional space over GF(3) has Hamming distance at most 1 to some code word. Using the code words as forecasts therefore guarantees 2nd prize. The football pool problem is to determine the minimal number A(k) of forecasts necessary to guarantee 2nd prize if k is the number of matches. Apparently the problem is solved for $k = \frac{1}{2}(3^r - 1)$. Besides these only A(2), A(3) and A(5) are known. The proof that A(5) = 27 takes 10 pages. (Cf. [7]). We will return to this problem in Section 5.1.

We have seen that the error-correcting capabilities of a linear code depend on the minimum weight of the code. If more is known about the weights of the code words we can provide information about the probability of error. For this purpose we introduce

(2.2.5) DEFINITION: If A_i denotes the number of code words of weight i in a code then

$$A(z) := \sum_{i=0}^{n} A_i z^i$$

is called the <u>weight enumerator</u> of the code; (n is the block length).

Consider a code with minimum distance $d \geq 2t + 1$, weight enumerator A(z) and suppose the decoding algorithm corrects up to t errors and otherwise fails. Since we can easily compute the probability that the error-pattern has weight $\leq t$ we can find the probability of a decoding error by computing the probability that the error pattern is in a coset with a leader of weight $\leq t$. If $\underline{x}$ is a codeword of weight i and we change ν of the 1's to 0's and μ of the 0's to 1's we obtain a vector of weight $i - \nu + \mu$ with distance $\nu + \mu$ to $\underline{x}$. Let $A_i(j,k)$ denote the number of words of weight k with distance j to a specific code word of weight i. We have

(2.2.6) $$\sum_j \sum_k A_i(j,k)\xi^j\eta^k = \sum_{\nu=0}^{i} \sum_{\mu=0}^{n-i} \binom{i}{\nu}\binom{n-i}{\mu}\xi^{\nu+\mu}\eta^{i-\nu+\mu} =$$

$$= (\xi+\eta)^i(1+\xi\eta)^{n-i}.$$

The probability that the error-pattern is in a coset with a leader of weight $\leq t$ is then (for a BSC with error probability p):

$$\sum_{j=0}^{t} \sum_{i=0}^{n} \sum_{k=0}^{n} A_i A_i(j,k) p^k (1-p)^{n-k}.$$

In order to calculate this probability we introduce the polynomial $P_t(x)$ of degree $\leq t$ by

$$P_t(x) := \sum_{j=0}^{t} \sum_{i=0}^{n} \sum_{k=0}^{n} A_i \cdot A_i(j,k) p^k (1-p)^{n-k} x^j.$$

Note that $P_t(x)$ is obtained by truncating $P_n(x)$. We have by (2.2.6)

(2.2.7) $$P_n(x) = \sum_{i=0}^{n} A_i (x-xp+p)^i (1-p+xp)^{n-i}$$

$$= (1-p+xp)^n \, A\left(\frac{x-xp+p}{1-p+xp}\right).$$

<u>Example</u>: Consider the BSC of Section 1.2 and suppose we use the (8,4) extended Hamming code (rate 1/2) with the incomplete decoding algorithm which corrects one error. By inspecting the generator matrix we immediately see that $A(z) = 1 + 14z^4 + z^8$. From (2.2.7) we find $P_8(x)$ and by truncating we find the required probability $P_1(1)$ that the decoding procedure does not fail. The result is

$$\text{Prob(correct decoding)} = (1-p)^8 + 8p(1-p)^7 = 1 - 28p^2 + 112p^3 - 210p^4 + \cdots .$$

$$\text{Prob(decoding)} = (1-p)^8 + 14(1-p)^4p^4 + p^8 + 8(1-p)^7p + 56(1-p)^3p^5 + 56p^3(1-p)^5 + 8p^7(1-p) =$$

$$= 1 - 28p^2 + 166p^3 - 476p^4 + \cdots .$$

$$\text{Prob(decoding error)} = 54p^3 - 266p^4 + \cdots = 0.00039,$$

i.e. about 0.04% while Prob (correct decoding) $\approx$ 0.9897

so that the other 1.03 % is practically completely decoding failure.

We shall now calculate $A(z)$ for the binary Hamming codes. If we take i - 1 columns of the parity check matrix H then there are 3 possibilities: (i) the sum of these columns can be 0, (ii) the sum of these columns can be one of the chosen columns, (iii) the sum of these columns can be equal to one of the remaining columns. The total number of possible choices is $\binom{n}{i-1}$. Possibility (i) can occur in A_{i-1} ways, possibility (ii) in $(n-i+2)A_{i-2}$ ways and possibility (iii) in iA_i ways. Therefore

$$iA_i = \binom{n}{i-1} - A_{i-1} - (n-i+2)A_{i-2} \,. \tag{2.2.8}$$

Obviously $A_{n-1} + A_n = 1$ and therefore (2.2.8) also holds for $i > n$. Multiply both sides of (2.2.8) by z^{i-1} and sum for $i = 1,2,\ldots,n+2$. The result is

$$A'(z) = (1+z)^n - A(z) - nzA(z) + z^2A'(z). \tag{2.2.9}$$

Since $A(0) = 1$ the solution of this linear differential equation is

$$A(z) = \frac{1}{n+1}(1+z)^n + \frac{n}{n+1}(1+z)^{\frac{n-1}{2}}(1-z)^{\frac{n+1}{2}} \,. \tag{2.2.10}$$

Let us now use this result to calculate the expected number of errors per block in a Hamming code of block length $n = 2^m - 1$. Look at the standard array described in Section 2.1. If an error pattern of weight i is a code word then none of the errors are corrected. If an error pattern of weight $i+1$ is under a code word of weight i then one error is corrected. If an error pattern of weight $i-1$ is under a code word of weight i then the decoding algorithm introduces an extra error! Therefore the expected number of errors is

$$E := \sum_{i=0}^{n} iA_i\left\{p^iq^{n-i} + (n-i)p^{i+1}q^{n-i-1} + ip^{i-1}q^{n-i+1}\right\} =$$

$$q^n\sum_{i=0}^{n} iA_i\left\{x^i + (n-i)x^{i+1} + ix^{i-1}\right\},$$

where $x := p/q$. Therefore $E = q^n\left\{((n-1)x^2+x+1)A'(x) + (x-x^3)A''(x)\right\}$. By (2.2.10) we then have

$$E = q^n\Big\{((n-1)x^2+x+1)\Big[\frac{n}{n+1}(1+x)^{n-1} - \frac{n}{n+1}(1+nx)(1+x)^{\frac{n-3}{2}}(1-x)^{\frac{n-1}{2}}\Big] + \tag{2.2.11}$$

$$+ (x-x^3)\Big[\frac{n(n-1)}{(n+1)}(1+x)^{n-2} + \frac{n(n-1)}{n+1}(nx^2+2x-1)(1+x)^{\frac{n-5}{2}}(1-x)^{\frac{n-3}{2}}\Big]\Big\}.$$

i.e.

$$E = \Big((n-1)\frac{p^2}{1-p} + 1\Big)\Big[\frac{n}{n+1}\Big(1 - (1 + (n-1)p)(1-2p)^{\frac{n-1}{2}}\Big)\Big] +$$

$$+ \frac{p(1-2p)}{1-p}\cdot\frac{n(n-1)}{n+1}\Big[1 + (np^2-3p^2+4p-1)(1-2p)^{\frac{n-3}{2}}\Big].$$

Now let us see how this expression behaves if we let n and p vary, $np = \alpha$ (fixed) and $n \to \infty$. Then

$$E \to [1 - (1+\alpha)e^{-\alpha}] + \alpha[1 - e^{-\alpha}] =$$

$$= \alpha + [1 - (1+2\alpha)e^{-\alpha}].$$

Now remark that α is the expected number of errors per block before decoding. If $\alpha > 1.2564\ldots$ the term in square brackets is positive, i.e. the expected number of errors per block is larger after decoding than before! With this example we hope to have made clear that it is useful to compute these error probabilities and also that we need codes which are much better than the Hamming codes.

<u>Examples</u>: For the BSC with $p = 0.02$ and the (7,4) Hamming code the expected number of errors per block before decoding is 0.14. After decoding this is 0.024 which is a reasonable improvement. For the (31,26) code these numbers are 0.62 and 0.407 which is already disappointing. For the (127,120) code we find 2.54 and 3.06, i.e. the code is useless!

If p is small the following estimate which is easier to obtain than (2.2.11) will be good enough for later applications. We consider the extended Hamming code of length $n = 2^m$. If an error pattern of weight $i \leq 1$ occurs the decoded word contains no errors, if $i = 2$ nothing happens and the decoded word contains 2 errors. In other cases it is possible that decoding increases the number of errors by one. Therefore the expected number of errors per block after decoding is at most

$$E^* := 2\binom{n}{2}p^2q^{n-2} + \sum_{i=3}^{n} (i+1)\binom{n}{i}p^iq^{n-i} =$$

(2.2.12) $$= 2\binom{n}{2}p^2\left[q^{n-2} + \sum_{i=3}^{n} \frac{(i+1)\binom{n}{i}}{2\binom{n}{2}} p^{i-2}q^{n-i}\right] \leq$$

$$\leq 2\binom{n}{2}p^2\left[q^{n-2} + \sum_{i=3}^{n} \binom{n-2}{i-2}p^{i-2}q^{n-i}\right] = n(n-1)p^2 < (np)^2.$$

This estimate will be used in Section 2.5.

2.3 Reed-Muller codes

Let $n = 2^m$ and let $\mathcal{R}^{(m)}$ be the m-dimensional vector space over GF(2) and $\mathcal{R}^{(n)}$ the n-dimensional space. Just as in Section 2.2 we consider an m by n matrix in which the columns are the binary representations of the integers $0, 1, \ldots, 2^{m-1}$. We now take the least significant bit first. The rows of this matrix are vectors $\underline{v}_1, \underline{v}_2, \ldots, \underline{v}_m$ in $\mathcal{R}^{(n)}$ and the columns are all the vectors of $\mathcal{R}^{(m)}$. If $\underline{u}_1, \underline{u}_2, \ldots, \underline{u}_m$ denote the columns with number 1, 2, 4, ..., 2^{m-1} (that is the standard basis of the space $\mathcal{R}^{(m)}$) then the vector in position $j = \sum_{i=1}^{m} \xi_{ij}2^{i-1}$ is $\underline{x}_j = \sum_{i=1}^{m} \xi_{ij}\underline{u}_i$.

In the space $\mathcal{R}^{(n)}$ we define a multiplication as follows:

(2.3.1) DEFINITION. If $\underline{a} = (a_0, a_1, \ldots, a_{n-1})$ and $\underline{b} = (b_0, b_1, \ldots, b_{n-1})$ then

$$\underline{ab} := (a_0b_0, a_1b_1, \ldots, a_{n-1}b_{n-1}).$$

Remark that for all $\underline{a} \in \mathcal{R}^{(n)}$ we have $\underline{a}^2 = \underline{a}$.

Let A_i be the subset of $\mathcal{R}^{(m)}$ consisting of those vectors which have a 1 in position i, i.e. $A_i := \{\underline{x}_j \in \mathcal{R}^{(m)} \mid \xi_{ij} = 1\}$. We can then interpret the vector $\underline{v}_i$ as the characteristic function of A_i. In this interpretation $\underline{v}_{i_1}\underline{v}_{i_2} \cdots \underline{v}_{i_k}$ is the characteristic function of $A_{i_1} \cap A_{i_2} \cap \cdots \cap A_{i_k}$. This last set is the subset of $\mathcal{R}^{(m)}$ consisting of the vectors which have 1's in all the positions $i_1, i_2, \ldots, i_k$.

This is an (m-k)-dimensional affine subspace of $\mathcal{R}^{(m)}$ and therefore consists of 2^{m-k} vectors of $\mathcal{R}^{(m)}$ (of course here we must assume that $i_1, i_2, \ldots, i_k$ are distinct). We have proved:

(2.3.2) LEMMA: $w(\underline{v}_{i_1} \underline{v}_{i_2} \cdots \underline{v}_{i_k}) = 2^{m-k}$.

In the following we shall need affine subspaces of $\mathcal{R}^{(m)}$ defined by fixing certain coordinates, the others being free. We can describe these using the A_i and their complements but to have a compacter notation we introduce

(2.3.3) DEFINITION: $C(i_1, i_2, \ldots, i_k)$:= set of all integers $j = \sum_{i=1}^{m} \xi_{ij} 2^{i-1}$ for which $\xi_{ij} = 0$ if $i \notin \{i_1, \ldots, i_k\}$. (The set $\{i_1, \ldots, i_k\}$ may also be empty).

As usual we denote $\{x + \alpha \mid x \in C\}$ by $C + \alpha$. With this notation we have for example:

$$\{\underline{x}_j \mid j \in C(2,3) + 8\} = \bar{A}_1 \cap A_4 \cap \bar{A}_5 \cap \bar{A}_6 \cap \cdots \cap \bar{A}_m.$$

The subset $\{\underline{x}_j\}$ of $\mathcal{R}^{(m)}$ has as characteristic function the vector $\underline{e}_j = (0,0,\ldots0,1,0,\ldots,0) \in \mathcal{R}^{(n)}$ with a 1 in position j, $(j = 0,1,\ldots,n-1)$. We introduce the vector $\underline{v}_0 := (1,1,\ldots,1)$, i.e. the characteristic function of $\mathcal{R}^{(m)}$. Then we have

$$\underline{e}_j = \prod_{i=1}^{m} \{\underline{v}_i + (1 + \xi_{ij})\underline{v}_0\}. \tag{2.3.4}$$

Each of the 2^m basis vectors $\underline{e}_j$ of $\mathcal{R}^{(n)}$ has thus been written as a product which, on expansion, yields a polynomial of degree $\leq m$ in the vectors $\underline{v}_1, \underline{v}_2, \ldots, \underline{v}_m$ (interpreting $\underline{v}_0$ as unity). This means that $\mathcal{R}^{(n)}$ is generated by the set of product vectors $\underline{v}_{i_1} \cdots \underline{v}_{i_k}$ $(k \leq m)$, i.e. the set $\{\underline{v}_0, \underline{v}_1, \ldots, \underline{v}_m, \underline{v}_1\underline{v}_2, \ldots, \underline{v}_1\underline{v}_2 \cdots \underline{v}_m\}$. Since this is a set of $1 + \binom{m}{1} + \cdots + \binom{m}{m} = 2^m = n$ vectors it is a basis of the vector space $\mathcal{R}^{(n)}$, i.e. the products are independent. We have proved:

(2.3.5) <u>THEOREM</u>: <u>The vectors</u> $\underline{v}_0, \underline{v}_1, \ldots, \underline{v}_m, \underline{v}_1\underline{v}_2, \underline{v}_1\underline{v}_3, \ldots, \underline{v}_{m-1}\underline{v}_m, \underline{v}_1\underline{v}_2\underline{v}_3, \ldots, \underline{v}_1\underline{v}_2 \cdots \underline{v}_m$ <u>form a basis of</u> $\mathcal{R}^{(n)}$.

Example: Let $m = 4$, $n = 16$. Then

$$
\begin{aligned}
\underline{v}_0 &= (\,1\,1\,1\,1\,1\,1\,1\,1\,1\,1\,1\,1\,1\,1\,1\,1\,) \\
\underline{v}_1 &= (\,0\,1\,0\,1\,0\,1\,0\,1\,0\,1\,0\,1\,0\,1\,0\,1\,) \\
\underline{v}_2 &= (\,0\,0\,1\,1\,0\,0\,1\,1\,0\,0\,1\,1\,0\,0\,1\,1\,) \\
\underline{v}_3 &= (\,0\,0\,0\,0\,1\,1\,1\,1\,0\,0\,0\,0\,1\,1\,1\,1\,) \\
\underline{v}_4 &= (\,0\,0\,0\,0\,0\,0\,0\,0\,1\,1\,1\,1\,1\,1\,1\,1\,) \\
\underline{v}_1\underline{v}_2 &= (\,0\,0\,0\,1\,0\,0\,0\,1\,0\,0\,0\,1\,0\,0\,0\,1\,) \\
\underline{v}_1\underline{v}_3 &= (\,0\,0\,0\,0\,0\,1\,0\,1\,0\,0\,0\,0\,0\,1\,0\,1\,) \\
\underline{v}_1\underline{v}_4 &= (\,0\,0\,0\,0\,0\,0\,0\,0\,0\,1\,0\,1\,0\,1\,0\,1\,) \\
\underline{v}_2\underline{v}_3 &= (\,0\,0\,0\,0\,0\,0\,1\,1\,0\,0\,0\,0\,0\,0\,1\,1\,) \\
\underline{v}_2\underline{v}_4 &= (\,0\,0\,0\,0\,0\,0\,0\,0\,0\,0\,1\,1\,0\,0\,1\,1\,) \\
\underline{v}_3\underline{v}_4 &= (\,0\,0\,0\,0\,0\,0\,0\,0\,0\,0\,0\,0\,1\,1\,1\,1\,) \\
\underline{v}_1\underline{v}_2\underline{v}_3 &= (\,0\,0\,0\,0\,0\,0\,0\,1\,0\,0\,0\,0\,0\,0\,0\,1\,) \\
\underline{v}_1\underline{v}_2\underline{v}_4 &= (\,0\,0\,0\,0\,0\,0\,0\,0\,0\,0\,0\,1\,0\,0\,0\,1\,) \\
\underline{v}_1\underline{v}_3\underline{v}_4 &= (\,0\,0\,0\,0\,0\,0\,0\,0\,0\,0\,0\,0\,0\,1\,0\,1\,) \\
\underline{v}_2\underline{v}_3\underline{v}_4 &= (\,0\,0\,0\,0\,0\,0\,0\,0\,0\,0\,0\,0\,0\,0\,1\,1\,) \\
\underline{v}_1\underline{v}_2\underline{v}_3\underline{v}_4 &= (\,0\,0\,0\,0\,0\,0\,0\,0\,0\,0\,0\,0\,0\,0\,0\,1\,)
\end{aligned}
$$

We can now define a number of linear codes which were discovered by D. E. Muller and for which I. S. Reed found a very nice decoding method.

(2.3.6) DEFINITION: The linear subspace of $\mathcal{R}^{(n)}$ which has as basis $\underline{v}_0, \underline{v}_1, \ldots, \underline{v}_m$ and all products $\underline{v}_{i_1}\underline{v}_{i_2} \cdots \underline{v}_{i_k}$ with $k \leq r$ is called the <u>r-th order Reed-Muller code</u> (RM-code) of length $n = 2^m$. The 0-th order RM code has $\underline{v}_0$ as a basis. It is a <u>repetition code</u> of length 2^m.

We remark that the following alternative description is equivalent: Consider all polynomials of degree $\leq r$ in m variables $x_1, x_2, \ldots, x_m$ which can take values in GF(2). Consider the possible set of values of $x_1, x_2, \ldots, x_m$:

$$x_1 = 0, 0, \ldots, 1$$
$$x_2 = 0, 0, \ldots, 1$$
- - - - - - - - - -
$$x_m = 0, 1, \ldots, 1$$

Successively substitute these sets of values in the polynomial $p(x_1,\ldots,x_m)$. Let the result be a code word. We then obtain the r-th order RM code of length 2^m. Note that if $\underline{a} = \underline{v}_{i_1}\underline{v}_{i_2} \cdots \underline{v}_{i_k}$ is a basis vector of the r-th order RM code and $\underline{b} = \underline{v}_{j_1}\underline{v}_{j_2} \cdots \underline{v}_{j_\ell}$ is a basis vector of the (m-r-1)-st order RM code then $\underline{a}\ \underline{b}$ is a basis vector of the (m-1)-st order RM code and by (2.3.2) $\underline{a}\ \underline{b}$ then has even weight which implies $(\underline{a},\underline{b}) = 0$. Now the dimension of the r-th order RM code is $1 + \binom{m}{1} + \cdots + \binom{m}{r}$ and the dimension of the (m-r-1)-st order RM code is $1 + \binom{m}{1} + \cdots + \binom{m}{m-r-1} = \binom{m}{m} + \binom{m}{m-1} + \cdots + \binom{m}{r+1}$. The sum of these two dimensions is $2^m = n$. Since we have just noted the orthogonality of respective basis vectors we have proved:

(2.3.7) <u>THEOREM</u>: <u>The dual of the</u> r-th <u>order</u> RM <u>code of length</u> 2^m <u>is the</u> (m-r-1)-st <u>order</u> RM <u>code of length</u> 2^m.

A special case is

(2.3.8) COROLLARY: The (m-2)-nd order RM code of length $n = 2^m$ is the $(2^m, 2^m-m-1)$ extended Hamming code.

Now we consider the problem of error correction when using the RM codes. First we shall find a way to express a vector $\underline{f} \in \mathcal{R}^{(n)}$ as a linear combination of products $\underline{v}_{i_1}\underline{v}_{i_2} \cdots \underline{v}_{i_k}$. From (2.3.4) it follows that $\underline{v}_{i_1}\underline{v}_{i_2} \cdots \underline{v}_{i_k}$ occurs in the expansion of $\underline{e}_j$ only if $\xi_{ij} = 0$ for all $i \notin \{i_1, i_2, \ldots, i_k\}$ and by (2.3.3) we can therefore write

$$(2.3.9) \qquad \underline{f} = \sum_{j=0}^{n-1} f_j \underline{e}_j = \sum_{(i_1,i_2,\ldots,i_k)} \Big(\sum_{j \in C(i_1,i_2,\ldots,i_k)} f_j \Big) \underline{v}_{i_1}\underline{v}_{i_2} \cdots \underline{v}_{i_k},$$

where the sum is over all k-tuples and $k = 0, 1, \ldots, m$. We shall use the r-th order RM code as follows. A sequence $(a_1,a_2,\ldots,a_M)$ of information bits (binary), with $M = 1 + \binom{m}{1} + \cdots + \binom{m}{r}$, is coded as

$$\underline{f} = a_1\underline{v}_0 + a_2\underline{v}_1 + \cdots + a_M\underline{v}_{m-r+1}\underline{v}_{m-r+2} \cdots \underline{v}_m =$$
$$= (f_0,f_1,\ldots,f_{n-1}) \in \mathcal{R}^{(n)}.$$

Let a_s be the coefficient of one of the products containing r factors, say $\underline{v}_{i_1}\underline{v}_{i_2} \cdots \underline{v}_{i_r}$. By (2.3.9) we have

(2.3.10) $$a_s = \sum_{j \in C(i_1,i_2,\ldots,i_r)} f_j \;.$$

Now take any $t \notin \{i_1,i_2,\ldots,i_r\}$. Then by (2.3.9) we know

(2.3.11) $$\sum_{j \in C(i_1,i_2,\ldots,i_r,t)} f_j = 0.$$

Since $C(i_1,i_2,\ldots,i_r,t) = C(i_1,i_2,\ldots,i_r) \cup [C(i_1,i_2,\ldots,i_r) + 2^{t-1}]$ and this is a unionof disjoint sets) we have from (2.3.10) and (2.3.11)

(2.3.12) $$a_s = \sum_{j \in C(i_1,i_2,\ldots,i_r)+2^{t-1}} f_j \quad ,(t \notin \{i_1,i_2,\ldots,i_r\}).$$

We can now consider the sum over $C(i_1,i_2,\ldots,i_r,t_1,t_2)$. We find a set of 2^{r+2} coefficients f_j summing to 0. This set contains the disjoint sets $C(i_1,i_2,\ldots,i_r)$, $C(i_1,i_2,\ldots,i_r) + 2^{t_1-1}$, $C(i_1,i_2,\ldots,i_r) + 2^{t_2-1}$ and for each of these the sum of the f_j's was a_s. Therefore this is true for the remaining set. By induction we then have:

(2.3.13) <u>THEOREM</u>: <u>For every information symbol</u> a_s <u>corresponding to a product of</u> r <u>vectors</u> $\underline{v}_i$ <u>we can split the set</u> $\{0,1,\ldots,n-1\}$ <u>into</u> 2^{m-r} <u>disjoint subsets</u> C <u>of</u> 2^r <u>elements such that for each of these</u> $a_s = \sum_{j \in C} f_j$.

To check this consider the example given above. Take $r = 2$ and let a_s be the coefficient of $\underline{v}_2\underline{v}_3$. The set $C(2,3)$ is $\{0,2,4,6\}$ by (2.3.3). Indeed, for each of the first 11 rows of the example except the row $\underline{v}_2\underline{v}_3$ the symbols in positions 0,2,4 and 6 sum to 0. The same holds for $C(2,3) + 8 = \{8,10,12,14\}$ and also for $\{1,3,5,7\}$ and $\{9,11,13,15\}$. So we have 4 "disjoint relations" involving the coefficient a_s of $\underline{v}_2\underline{v}_3$.

Now let us assume the received message is $\underline{x} = (x_0,x_1,\dots,x_{n-1})$. Consider as we did above an information symbol a_s corresponding to a product of r of the $\underline{v}_i$'s. In the absence of errors we have 2^{m-r} disjoint relations $a_s = \sum_{j\in C_\nu} x_j$ $(\nu = 1,2,\dots,2^{m-r})$. If the vector $\underline{x}$ contains less than $\frac{1}{2}\cdot 2^{m-r}$ errors then the majority of the relations will still hold. Hence we can then determine a_s by a majority decision. After doing this for all information symbols corresponding to products of r of the v_i's we subtract the linear combination of the $\underline{v}_{i_1} \cdots \underline{v}_{i_r}$ that we have found. The transmitted and the received message have been reduced to the (r-1)-st order RM code and we repeat the procedure. This shows that the code can correct $\leq 2^{m-r-1} - 1$ errors and therefore has minimum distance $\geq 2^{m-r} - 1$. Since all vectors in the code have even weight and the basis vectors $\underline{v}_{i_1} \cdots \underline{v}_{i_r}$ have weight 2^{m-r} this is the minimum weight of the code:

(2.3.14) <u>THEOREM</u>: <u>The r-th order RM code has minimum weight</u> 2^{m-r}.

We remark that several of the proofs given in this section could have been replaced by more geometrical arguments. The interested reader can construct such proofs as an exercise.

2.4. Threshold decoding

The method of decoding used in Section 2.3 is a variation of a decoding method which is used for many different codes and which is known as <u>threshold decoding</u>. A very simple example will serve as an introduction.

Consider the binary repetition code of length $2n + 1$. This code consists of the all-zero word and the all-one word. As a system of parity-check equations we can choose

$$(2.4.1) \qquad \begin{cases} x_1 + x_2 = 0, \\ x_1 + x_3 = 0, \\ \text{--------} \\ x_1 + x_{2n+1} = 0. \end{cases}$$

If the received word is $\underline{y} = \underline{x} + \underline{e}$ then $y_1 + y_i = e_1 + e_i$ $(i = 2,\ldots,2n+1)$. If more than n of the expressions $y_1 + y_i$ is 1 this can be explained by either less than n errors among which $e_1 = 1$ or by more than n errors and $e_1 = 0$. The former is more likely. A majority vote among the $y_1 + y_i$ decides whether $e_1 = 0$ or 1. If the number of votes exceeds the threshold n then we set $e_1 = 1$.

The equations (2.4.1) are a special example of the following situation:

(2.4.2) DEFINITION: Let V be a linear code with block length n. A set of parity check equations for V is said to be orthogonal on the set of positions $P \subset \{1,2,\ldots,n\}$ iff

(a) for every $i \in P$ the term x_i occurs in each equation with a non-zero coefficient;

(b) for every $i \notin P$ the term x_i occurs in at most one of the equations with a non-zero coefficient.

Let us now demonstrate how to use such a set of parity-check equations. Consider the dual of the (15,11) Hamming code. If H is the parity-check equation of this Hamming code then for every pair of distinct columns of H there is a third column of H such that the 3 columns sum to zero. Each of the weight-3 code words of the Hamming code obtained in this way gives us a parity-check equation for the dual code with three terms. Among these the set of 7 equations:

$$(2.4.3) \qquad \begin{cases} x_1 = x_2 + x_3, \\ x_1 = x_4 + x_5, \\ \text{---------} \\ x_1 = x_{14} + x_{15} \end{cases}$$

is orthogonal on position number 1. For the symbol x_1 we therefore have the equations:

$$(2.4.4.)\quad \begin{cases} y_1 = x_1 + e_1 , \\ y_2 + y_3 = x_1 + (e_2+e_3), \\ \text{-------------------} \\ y_{14} + y_{15} = x_1 + (e_{14}+e_{15}). \end{cases}$$

The code under consideration has minimum distance 8. If the number of errors in the received message $\underline{y}$ is less than 4 the majority of the left-hand sides of (2.4.4) will have the value x_1. Once again a majority vote with threshold 4 decides the value x_1. If the vote comes out four to four then 4 errors are detected. Now remark that for a BSC with error probability p we have Prob $(e_1 = 0) = 1 - p$ and Prob $(e_i + e_j) = 0 = (1-p)^2 + p^2 < 1 - p$. Therefore in the case of a tie it is more likely that the estimate $y_1 = x_1$ is correct and we decode in this way. After decoding the first symbol we proceed in the same way, using a set of parity-check equations orthogonal on the second position, etc. Not only are threshold decoders often easily implemented but they also have the additional feature of correcting many more error-patterns than the code is designed for. For example in the code treated above many patterns of 4 errors are correctly decoded, e.g. (111100000000000).

By treating one example we shall show that threshold decoding can also be used for RM codes. Let us consider the (32,16)-second order RM code. By (2.3.14) this is a 3-error-correcting code and by (2.3.7) the code is self-dual. Hence every parity-check equation for the code contains at least 8 terms. It is therefore impossible to find a set of 6 parity-check equations orthogonal on one position. For a majority decision in case of 3 errors we need at least 6 equations! The solution for this problem is obtained by making the decision in two steps. First we need to know more about the parity-check equations of a RM code. The following theorem is useful:

(2.4.5) THEOREM: Let V be the (m-k)-th order RM code of length 2^m. Then the characteristic function of any k-dimensional affine subspace of $\mathcal{R}^{(m)}$ is a code word of V.

Proof: Consider a k-dimensional affine subspace A of $\mathcal{R}^{(m)}$ and let $\underline{f} \in \mathcal{R}^{(n)}$ be the characteristic function of A. Take $r > m - k$ and consider the coefficient of $\underline{v}_{i_1}\underline{v}_{i_2} \cdots \underline{v}_{i_r}$ in the expansion of $\underline{f}$ as given by (2.3.9). This coefficient is $\sum_{j \in C(i_1, i_2, \ldots, i_r)} f_j$ which is just the number of vectors in the affine subspace A which are also in the linear subspace $L := \{\underline{x}_j \in \mathcal{R}^{(m)} \mid j \in C(i_1, i_2, \ldots, i_r)\}$. Since L has dimension $r > m - k$ the intersection of L and A is empty or an affine subspace of dimension > 0. This intersection therefore has an even number of points, i.e. the coefficient of $\underline{v}_{i_1}\underline{v}_{i_2} \cdots \underline{v}_{i_r}$ is 0. In other words, $\underline{f}$ is in the (m-k)-th order RM code.

A consequence of (2.4.5) is that every 3-dimensional affine subspace of $\mathcal{R}^{(5)}$ provides us with a parity check for the 2-nd order RM code of length 32 since this code is self-dual. In the following array (a,b,c,d) will denote a set of positions such that $\{\underline{x}_a, \underline{x}_b, \underline{x}_c, \underline{x}_d\}$ is a 2-dimensional affine subspace of $\mathcal{R}^{(5)}$. In every row these 2-dimensional subspaces are translates of the first space in the row, i.e. the union of such a pair forms a set of positions of a parity-check equation of the RM code! The subspaces in the first column form a set of positions which in a sense analogous to (2.4.2) is orthogonal on position 0.

0,1,2,3	4,5,6,7	8,9,10,11	12,13,14,15	16,17,18,19	20,21,22,23	24,25,26,27	28,29,30,31
0,4,8,12	1,5,9,13	2,6,10,14	3,7,11,15	16,20,24,28	17,21,25,29	18,22,26,30	19,23,27,31
0,5,10,15	1,4,11,14	2,7,8,13	3,6,9,12	16,21,26,31	17,20,27,30	18,23,24,29	19,22,25,28
0,6,16,22	1,7,17,23	2,4,18,20	3,5,19,21	8,14,24,30	9,15,25,31	10,12,26,28	11,13,27,29
0,7,19,20	1,6,18,21	2,5,17,22	3,4,16,23	8,15,27,28	9,14,26,29	10,13,25,30	11,12,24,31
0,9,18,27	1,8,19,26	2,11,16,25	3,10,17,24	4,13,22,31	5,12,23,30	6,15,20,29	7,14,21,28

Once again let a code word $\underline{x}$ be transmitted and let $\underline{y} = \underline{x} + \underline{e}$ be received. We have

$$(2.4.6)\quad \begin{cases} y_0+y_1+y_2+y_3+y_4+y_5+y_6+y_7 = (e_0+e_1+e_2+e_3) + (e_4+e_5+e_6+e_7), \\ \text{- -} \\ y_0+y_1+y_2+y_3+y_{28}+y_{29}+y_{30}+y_{31} = (e_0+e_1+e_2+e_3) + (e_{28}+e_{29}+e_{30}+e_{31}). \end{cases}$$

Just as in (2.4.4) a majority vote among the left-hand sides of (2.4.6) decides whether $e_0+e_1+e_2+e_3$ is 0 or 1. The threshold is taken as $3\frac{1}{2}$, i.e. $e_0+e_1+e_2+e_3$ is taken to be 1 if 4 or more of the syndrome elements are 1. If the number of errors is 3 or less, then $e_0+e_1+e_2+e_3$ is given the correct value. We repeat this procedure for each row of the array and then let a final majority vote over the first column with threshold $3\frac{1}{2}$ decide whether $e_0 = 0$ or $e_0 = 1$. This procedure is applied for every e_i. It is easily seen that if the number of errors is 3 or less this decoding procedure reproduces $\underline{x}$. There are many error patterns of weight > 3 which are also decoded correctly. This example of two-step threshold decoding gives an idea of how the method works in general. For more information on this subject we refer the reader to J. L. Massey, Threshold Decoding, MIT-Press, 1963.

2.5. Direct-product codes

Consider a block code with length $n = n_1n_2$. Instead of writing the code words as row vectors of length n we can represent the code words by matrices with n_1 rows and n_2 columns. One way of doing this is e.g. representing the code word $\underline{a} = (a_0,a_1,\ldots,a_{n-1})$ by the matrix $A := [a_{ij}]$, $(i = 0,1,\ldots,n_1-1;\ j = 0,1,\ldots,n_2-1)$, where $a_{ij} := a_{in_2+j}$. This is called the canonical ordering.

(2.5.1) DEFINITION: Let V_1 be a linear code of length n_1 and V_2 a linear code of length n_2. Let V be a code of length n_1n_2 represented by n_1 by n_2 matrices (with the canonical ordering). We shall say that V is the <u>direct product</u> of V_1 and V_2 iff V consists of all code words for which the matrix representation has the following properties: (i) each column of a matrix is (the transpose of) a code word of V_1, (ii) each row of a matrix is a code word of V_2.

It is clear that the direct product of two linear codes is again a linear code. If we interchange the factors V_1 and V_2 we obtain an equivalent code in the sense defined in Section 2.1. We shall denote the direct product of V_1 and V_2 by $V_1 \times V_2$.

(2.5.2) THEOREM: The minimum weight of $V_1 \times V_2$ is the product of the minimum weights of V_1 and V_2.

Proof: Let V_1 have length n_1 and let V_2 have length n_2 and let the n_1 by n_2 matrix A represent a non-zero code word of $V := V_1 \times V_2$. (We shall see below that there are such words). At least one row of A contains an element $\neq 0$ and hence at least $w(V_2)$ non-zero elements where $w(V_2)$ denotes the minimum weight of V_2. Each of these non-zero elements is in a column with at least $w(V_1)$ non-zero elements, i.e. at least $w(V_1)w(V_2)$ elements of A are not zero.

Let V_1 be an (n_1,k_1) linear code and V_2 an (n_2,k_2) linear code. Let G_1 and G_2 be the generator matrices of these codes, both in reduced echelon form. We denote the row vectors of G_1 by $\underline{g}_i^{(1)}$ $(i = 0,\dots,k_1-1)$ and the row vectors of G_2 by $\underline{g}_j^{(2)}$ $(j = 0,\dots,k_2-1)$. Now define the matrix A_{ij} $(0 \leq i < k_1,\ 0 \leq j < k_2)$ as follows: The first k_1 rows of A_{ij} are zero except the i-th row which is $\underline{g}_j^{(2)}$. The first k_2 columns of A_{ij} are zero except the j-th column which is $\underline{g}_i^{(1)T}$. For $k \geq k_1$ the k-th row is $g_{ik}^{(1)}\ \underline{g}_j^{(2)}$. Consequently for $\ell \geq k_2$ the ℓ-th column is $g_{j\ell}^{(2)}\ \underline{g}_i^{(1)T}$. This matrix has the form

$$A_{ij} = \left(\begin{array}{ccc|c} 0 & \begin{matrix}0\\ \vdots\\ 0\end{matrix} & 0 & 0 \\ 0 \cdots 0 & 1 & 0 \cdots 0 & ** \cdots ** \\ 0 & \begin{matrix}0\\ \vdots\\ 0\end{matrix} & 0 & 0 \\ \hline 0 & \begin{matrix}*\\ \vdots\\ *\end{matrix} & 0 & * \end{array}\right).$$

Obviously this matrix represents a code word of $V_1 \times V_2$ and each code word of $V_1 \times V_2$ must be represented by a linear combination of such matrices. This proves:

(2.5.3) THEOREM: If V_1 is an (n_1,k_1) linear code and V_2 is an (n_2,k_2) linear code then $V_1 \times V_2$ is an (n_1n_2, k_1k_2) linear code.

In the representation used in the proof of (2.5.3) the elements in the $k_1 \times k_2$ submatrix of a matrix A are the information symbols. We now shall try to find the generator matrix of $V_1 \times V_2$. We need a definition from matrix theory:

(2.5.4) DEFINITION: If $A := [a_{ij}]$ is an n_1 by m_1 matrix and B is an n_2 by m_2 matrix then the Kronecker product $A \times B$ is the matrix

$$A \times B := \begin{pmatrix} a_{11}B & a_{12}B & \cdots & a_{1m_1}B \\ \hline a_{n_11}B & & \cdots & a_{n_1m_1}B \end{pmatrix}.$$

Example:

$$\begin{pmatrix} 2 & 1 \\ 0 & -1 \end{pmatrix} \times \begin{pmatrix} 3 & 1 \\ 1 & 2 \end{pmatrix} = \begin{pmatrix} 6 & 2 & 3 & 1 \\ 2 & 4 & 1 & 2 \\ 0 & 0 & -3 & -1 \\ 0 & 0 & -1 & -2 \end{pmatrix}.$$

Now consider $G_1 \times G_2$. If we take the (k_2i+j)-th row of this matrix and represent it in its canonical form as an n_1 by n_2 matrix then this matrix is the matrix A_{ij} defined above. Hence $G_1 \times G_2$ is the generator matrix of $V_1 \times V_2$. For this reason direct-product codes are also called Kronecker product codes.

The following decoding algorithm is used for direct-product codes. If a matrix A representing a code word is received then first the rows of A are decoded by the procedure for the code V_2 and then the columns of A are decoded by the procedure for V_1. Due to the simplicity of this rule implementation is not too difficult. Notice that if a burst-error occurs some of the rows of A will contain very many errors but the burst will not seriously effect the columns and it is possible that it is completely corrected in the second phase. An even more effective way of combating burst errors is possible if $(n_1,n_2) = 1$. The matrix $A = [a_{ij}]$, $i = 0,\ldots,n_1-1$,

$j = 0,\dots,n_2-1$ is transmitted as the sequence $\underline{c} := (c_0,c_1,\dots,c_{n_1n_2-1})$ where $c_k := a_{ij}$ if $k \equiv i \pmod{n_1}$ and $k \equiv j \pmod{n_2}$. By the Chinese remainder theorem this is a one-one correspondence between the elements of A and $\underline{c}$. Using this cyclic ordering a burst error in $\underline{c}$ will be distributed among the rows and columns of A. It would be very useful if a not too complicated decoding algorithm for product codes could be found which actually corrects the error patterns that one knows can be corrected. The following example shows that this is not true for the procedure described above: If V_1 and V_2 both have minimum distance 3 e.g. if they are Hamming codes then by (2.5.2) $V_1 \times V_2$ is a 4-error-correcting code. If a word of $V_1 \times V_2$ is received with 4 errors in places which in the matrix representation are in rows 0 and 1 and columns 0 and 1 then these 4 errors are not corrected, in fact more errors are introduced in both phases of the procedure described above!

Recall that Shannon's theorem, proved in Section 1.4, asserts that for a BSC a sequence of codes can be found such that the corresponding sequence of error probabilities tends to 0 while the rates remain $> R$ if $R < C$ (C is the capacity of the channel). By using product codes, P. Elias constructed a sequence of codes for which the error probabilities tend to 0 and the rates are bounded away from zero. Although this is not quite as good as is possible according to Shannon's theorem it is still the only known sequence of codes with this property! The construction runs as follows:

Consider a BSC with error probability p and assume p is small enough for there to be an extended Hamming code of length 2^m for which the expected number of errors per block at the receiver is $< \frac{1}{2}$. We take this extended Hamming code as the first code in our sequence, say V_1. If code V_i has been defined then code V_{i+1} is obtained by taking the direct product of the extended Hamming code of length 2^{m+i} and V_i. Denote the length of V_i by n_i and the dimension by k_i. Furthermore let n_ip_i be the expected number of errors per block after decoding of V_i. We use the decoding procedure described above. In every phase of decoding the errors in a block are in different words which were corrected in the previous phase and therefore these errors are independent. We therefore have by (2.5.3) and (2.2.12):

(2.5.5) $$n_i = \prod_{j=0}^{i-1} 2^{m+j} = 2^{mi+\frac{1}{2}i(i-1)},$$

(2.5.6) $$k_i = \prod_{j=0}^{i-1} (2^{m+j} - m - j - 1),$$

(2.5.7) $$p_i = p^{2^i} \prod_{j=0}^{i-1} (2^{m+i-j-1})^{2^j} .$$

Note that the rate of V_i is

$$k_i/n_i = \prod_{j=0}^{i-1} \left(1 - \frac{m+j+1}{2^{m+j}}\right) > \prod_{j=0}^{\infty} \left(1 - \frac{m+j+1}{2^{m+j}}\right) > 0.$$

Furthermore by (2.5.7) since $2^m p < \frac{1}{2}$:

$$p_i = p^{2^i}\, 2^{(m+1)2^i-m-i-1} < (2^{m+1}p)^{2^i} \to 0 \quad \text{if } i \to \infty,$$

and also $n_i p_i \to 0$ if $i \to \infty$. (In fact the probability that a decoded word contains an error tends to 0). This shows that the codes V_i, the <u>Elias codes</u>, have the property stated above.

2.6 <u>Problems</u>

(2.6.1) Let $H := \begin{pmatrix} 1&1&0&1&0&1 \\ 1&1&0&0&1&0 \\ 1&0&1&1&0&0 \end{pmatrix}$ be the parity-check matrix of a binary linear code. Decode the following received messages: (a) 110110, (b) 010100.

(2.6.2) If an (n,k) linear code over GF(q) has a generator G in which no column with only zeros occurs then the sum of the weights of the code words is $n(q-1)q^{k-1}$. Prove this.

(2.6.3) If V is a binary linear (n,k) code then all code words have even weight or the code words of even weight form an (n,k-1) linear code. Prove this. What is the relation between the parity-check matrices of the two codes?

(2.6.4) If a (22,14) 2-error-correcting code exists what can be said about the weights of the coset leaders in the standard array?

(2.6.5) Let p be a prime. Is there always an (8,4) self-dual linear code over GF(p)?

(2.6.6) Determine the weight enumerator of the extended binary Hamming code.

(2.6.7) Determine the weight enumerator of a Hamming code over GF(q) if $q > 2$.

(2.6.8) Prove that the row vectors $\underline{v}_0, \underline{v}_1, \ldots, \underline{v}_1\underline{v}_2 \cdots \underline{v}_m$ as described in Section 2.3 can be rearranged in such a way that they form a matrix with zeros below the main diagonal.

(2.6.9) Consider the 2nd-order RM code of length 32. What is the rate of this code? Decode the following received word: (10110100101101001011000000001111).

(2.6.10) Determine the number of words of weight 4 in the 2nd-order RM code of length 16.

(2.6.11) Analyze the possible outcomes of the decoding procedure described in Section 2.4 in the case of the 2nd-order RM code of length 16 if the error pattern has weight 4.

III. CYCLIC CODES

3.1 Introduction

In this chapter $\mathcal{R}^{(n)}$ will denote the n-dimensional vector space over GF(q). We shall make the restriction $(n,q) = 1$. Consider the ring R of all polynomials with coefficients in GF(q), i.e. (GF(q)[x],+,). Let S be the principal ideal in R generated by the polynomial $x^n - 1$, i.e. $S := ((\{x^n-1\}),+,\ )$. R/S is the residue class ring R mod S, i.e. $(GF(q)[x] \bmod (\{x^n-1\}),+,\ )$. The elements of this ring can be represented by polynomials of degree $< n$ with coefficients in GF(q). The additive group of R/S is isomorphic to $\mathcal{R}^{(n)}$. An isomorphism is given by associating the vector $\underline{a} = (a_0,a_1,\dots,a_{n-1})$ with the polynomial $a_0 + a_1x + \cdots + a_{n-1}x^{n-1}$.

From now on we shall make use of this isomorphism in the following way: The elements of the vector space $\mathcal{R}^{(n)}$ are either referred to as "words" (or vectors) or as "polynomials" and generally we do not distinguish between the two representations. In addition to the vector space structure there is now also a multiplication in $\mathcal{R}^{(n)}$ which is just the multiplication of polynomials mod $(x^n - 1)$.

Notice that the multiplication by x amounts to applying the cyclic shift which sends $(a_0,a_1,\dots,a_{n-1})$ into $(a_{n-1},a_0,a_1,\dots,a_{n-2})$.

(3.1.1) DEFINITION: A k-dimensional linear subspace V of $\mathcal{R}^{(n)}$ is called a <u>cyclic code</u> (or cyclic subspace) if

$$\forall_{(a_0,\dots,a_{n-1})\in \mathcal{R}^{(n)}} \left[(a_0,a_1,\dots,a_{n-1}) \in V \Rightarrow (a_{n-1},a_0,a_1,\dots,a_{n-2}) \in V\right].$$

Since we are using two terminologies simultaneously we can interpret V as a subset of the ring R/S.

(3.1.2) <u>THEOREM</u>: <u>V is a cyclic code iff V is an ideal in</u> R/S.

<u>Proof</u>: (i) If V is an ideal in R/S and $(a_0,a_1,\dots,a_{n-1}) \in V$ then the polynomial obtained by multiplying by x is also in V. This is the cyclic shift

$(a_{n-1},a_0,\dots,a_{n-2})$. (ii) If $(a_0,a_1,\dots,a_{n-1}) \in V$ implies $(a_{n-1},a_0,\dots,a_{n-2}) \in V$ then for every polynomial $a(x) \in V$ we have $xa(x) \in V$. But then also $x^2a(x) \in V$, $x^3a(x) \in V$ etc. Hence we also have $p(x)a(x) \in V$ for any polynomial $p(x)$, i.e. V is an ideal in R/S.

From now on we shall write $\mathcal{R}$ instead of $\mathcal{R}^{(n)}$ or R/S and denote the ideal generated by a polynomial $g(x)$ by $\mathcal{R}g(x)$. Every ideal in $\mathcal{R}$ is a principal ideal generated by the monic polynomial of lowest degree in the ideal, say $g(x)$, where $g(x)$ is a divisor of $x^n - 1$ in R. We shall call $g(x)$ the <u>generator</u> (-polynomial) of the cyclic code.

Let $x^n - 1 = f_1(x)f_2(x) \cdots f_t(x)$ be the decomposition of $x^n - 1$ into irreducible factors in R. Since $(n,q) = 1$ there are no multiple factors.

(3.1.3) DEFINITION: The cyclic code generated by $f_i(x)$ is called a <u>maximal cyclic code</u> and denoted by $M_i^+ := \mathcal{R}\, f_i(x)$.

Note that a maximal cyclic code is a maximal ideal in $\mathcal{R}$.

Let π_j denote the permutation $k \to jk \pmod n$ of the set of integers $\{0,1,\dots,n-1\}$. We can also interpret π_j as an automorphism of the ring $\mathcal{R}$ by taking the natural definition $\pi_j(\sum_{i=0}^{n-1} a_i x^i) := \sum_{i=0}^{n-1} a_i x^{ji}$.

Of special interest to us is the binary case ($q = 2$, n odd). For this case let $C_1, C_2, \dots, C_t$ denote the cycles of the permutation π_2. E.g. if $n = 63$ we have the cycles (1,2,4,8,16,32), (3,6,12,24,48,33), (5,10,20,40,17,34), (7,14,28,56,49,35), (9,18,36), (11,22,44,25,50,37), (13,26,52,41,19,38), (15,30,60,57,51,39), (21,42), (23,46,29,58,53,43), (27,54,45), (31,62,61,59,55,47), (0).

From the theory of finite fields we know that if α is a primitive n-th root of unity (in a suitable extension field of GF(2)) the polynomial $\prod_{i \in C} (x-\alpha^i)$ where C is any of the cycles mentioned above is an irreducible factor of $x^n - 1$ in R. Hence to each of the t cycles corresponds one of the factors f_i and one maximal cyclic code. In the example considered above we can take α to be a primitive element of $GF(2^6)$. In this case the polynomial $(x-\alpha^9)(x-\alpha^{18})(x-\alpha^{36})$ is an irreducible factor of $x^7 - 1$ (i.e. α^9 is a primitive element of $GF(8)$). We say e is the exponent of $a(x)$ if e is the least positive integer such that $a(x)$ divides $x^e - 1$ (in R). For any cycle C the exponent of the corresponding polynomial is n/d if d is the greatest common divisor of the numbers in the cycle.

Remark: If $a(x)$ is a divisor of $x^n - 1$ with exponent $e < n$, i.e. $a(x)|x^e - 1$ (in R) then $x^e - 1$ is a code word of weight 2 in the code generated by $a(x)$. It is clear that we shall have little interest in such codes for practical purposes!

We can now find all cyclic codes of length n over $GF(q)$ by factoring $x^n - 1$ into $f_1(x) \cdots f_t(x)$ and taking any of the 2^t factors of $x^n - 1$ as a generator. Many of these codes will be equivalent!

Let $g(x)$ be the generator of a cyclic code of length n over $GF(q)$ and let $x^n - 1 = g(x)h(x)$ (in R). If g has degree $n - d$ then h has degree d. (We observe that the polynomials $g(x), xg(x), \ldots, x^{d-1}g(x)$ form an independent set, in fact a basis, in the vector space $\mathcal{R}$. Therefore, if $g(x) = g_0 + g_1x + \cdots + g_{n-d}x^{n-d}$ then a generator matrix of the cyclic code $\mathcal{R}\, g(x)$ is

$$G := \begin{pmatrix} g_0 g_1 & \cdots\cdots g_{n-d} & 0\ 0 \cdots\cdots\cdots 0 \\ 0\ g_0 g_1 & \cdots\cdots\ g_{n-d} & 0 \cdots\cdots 0 \\ -------- & & \\ 0\ 0 & \cdots\cdots 0\ g_0 & \cdots\cdots\cdots\cdots g_{n-d} \end{pmatrix}.$$

A more convenient form of the generator matrix is obtained by defining, (for $i \geq n - d$), $x^i =: g(x)q_i(x) + r_i(x)$ where $r_i(x)$ is a polynomial of degree $< n - d$. The polynomials $x^i - r_i(x)$ are code words of $\mathcal{R}\, g(x)$ and form a basis. Taking these basis vectors the generator matrix has the form $(-R\ I)$ where I is the identity matrix.

Let $h(x) = h_0 + h_1x + \cdots + h_dx^d$. In $\mathcal{R}$ we have $g(x)h(x) = 0$, i.e. for $i = 0,1,\ldots,n-1$ we have

$$g_0h_i + g_1h_{i-1} + \cdots + g_{n-d}h_{i-n+d} = 0$$

(where the subscripts are to be interpreted mod n). Therefore

$$H := \begin{pmatrix} 0 & 0 & \cdots\cdots & 0 & h_d & \cdots\cdots & h_1 & h_0 \\ 0 & \cdots\cdots & 0 & h_d & \cdots\cdots & & h_0 & 0 \\ \hline h_d & h_{d-1} & \cdots & h_0 & 0 & \cdots\cdots & & 0 \end{pmatrix}$$

is a parity check matrix of the code. Notice that the code $\mathcal{R}h(x)$ is equivalent to the dual of $\mathcal{R}g(x)$ (by the permutation $x \to x^{-1}$). Because of this equivalence we shall from now on refer to the code $\mathcal{R}h(x)$ as the dual of the code $\mathcal{R}g(x)$. The polynomial $h(x)$ is called the <u>check-polynomial</u> of the code $\mathcal{R}g(x)$.

An ideal V in $\mathcal{R}$ is called a <u>minimal ideal</u> if it contains no subideal other than $\{0\}$. From the description of the ideals in $\mathcal{R}$ given above we see that the minimal ideals are the duals of the maximal ideals. We denote by M_i^- the dual of M_i^+. The generator of M_i^- is $(x^n-1)/f_i(x)$. The codes M_i^- are called <u>irreducible cyclic codes</u>.

(3.1.4) DEFINITION: If V_1 and V_2 are ideals in $\mathcal{R}$ we denote by $V_1 + V_2$ the ideal generated by V_1 and V_2, i.e. the smallest ideal which contains V_1 and V_2.

Notice that $V_1 \cap V_2$ is also an ideal in $\mathcal{R}$.

(3.1.5) <u>THEOREM</u>: <u>If</u> $V_1 := \mathcal{R}g_1(x)$ <u>and</u> $V_2 := \mathcal{R}g_2(x)$ <u>then</u>

(i) $V_1 \cap V_2$ <u>is generated by the least common multiple of</u> $g_1(x)$ <u>and</u> $g_2(x)$,

(ii) $V_1 + V_2$ <u>is generated by the greatest common divisor of</u> $g_1(x)$ <u>and</u> $g_2(x)$.

We leave the proof to the reader as an exercise.

We remark that $M_i^- \cap M_j^- = \{\underline{0}\}$ if $i \neq j$ and that $M_i^+ \cap M_j^+ = \mathcal{R}(f_i(x)f_j(x))$. From (3.15) it follows that an ideal V in $\mathcal{R}$ is the "sum" of the minimal ideals contained in V.

To conclude this introduction we treat a simple example namely $n = 7$, $q = 2$. We have $x^7 - 1 = (x-1)(x^3+x^2+1)(x^3+x+1)$. If we take $g(x) = x^3 + x + 1$ we find the maximal cyclic code with generator

$$G = \begin{pmatrix} 1\,1\,0\,1\,0\,0\,0 \\ 0\,1\,1\,0\,1\,0\,0 \\ 0\,0\,1\,1\,0\,1\,0 \\ 0\,0\,0\,1\,1\,0\,1 \end{pmatrix} .$$

Since $h(x) = x^4 + x^2 + x + 1$ we have

$$H = \begin{pmatrix} 0\,0\,1\,0\,1\,1\,1 \\ 0\,1\,0\,1\,1\,1\,0 \\ 1\,0\,1\,1\,1\,0\,0 \end{pmatrix} .$$

It is not hard to see that this code is equivalent to the (7,4) Hamming code. (See 3.2.1).

3.2 The zeros of a cyclic code

If $g(x)$ is the generator of a cyclic code and $\alpha_1, \alpha_2, \ldots, \alpha_{n-d}$ are the zeros of g in a suitable extension field of GF(q) then a polynomial $a(x)$ is a code word of $\mathcal{R}g(x)$ iff $a(\alpha_1) = a(\alpha_2) = \cdots = a(\alpha_{n-d}) = 0$. In this way we have an alternate description of cyclic codes namely by the zeros common to all code words. It is no necessary to give all the zeros of $g(x)$ to specify $g(x)$. If, in the notation of Section 3.1, $g(x) = f_{i_1}(x)f_{i_2}(x) \cdots f_{i_k}(x)$ it is sufficient to require that $a(\alpha_{i_1}) = a(\alpha_{i_2}) = \cdots = a(\alpha_{i_k}) = 0$ where α_{i_j} is a zero of f_{i_j}. From the theory of finite fields we know that if α is a zero of a polynomial with coefficients in GF(q) then α^q is also a zero of this polynomial.

Let us now consider the simplest example namely the case where only one zero is required. We take $n = 2^m - 1$, $q = 2$. Let α be a primitive element of $GF(2^m)$ and $m_1(x) = (x-\alpha)(x-\alpha^2)(x-\alpha^4) \cdots (x-\alpha^{2^{m-1}})$ the minimal polynomial of α. (In the following we shall always denote the minimal polynomial of α^i where α is a primitive element of $GF(2^m)$ by $m_i(x)$. In these cases m is fixed). Every element of $GF(2^m)$ can be expressed uniquely as $\sum_{i=0}^{m-1} \epsilon_i\alpha^i$ where $\epsilon_i \in GF(2)$. Let H be the matrix for which the j-th column $(j = 0,1,\ldots,2^m-2)$ is $(\epsilon_0,\epsilon_1,\ldots,\epsilon_{m-1})^T$ of $\alpha^j = \sum_{i=0}^{m-1} \epsilon_i\alpha^i$.

We have seen that $a(x) = a_0 + a_1x + \cdots a_{n-1}x^{n-1} \in \mathcal{R}m_1(x)$ iff $a(\alpha) = 0$ but this is so iff $\underline{a}H^T = \underline{0}$. Therefore H is a parity-check matrix of the cyclic code generated by $m_1(x)$. Since H is obtained by permuting the binary representations of 1, 2, ..., 2^m-1 we have proved:

(3.2.1) THEOREM. The binary cyclic code of length $n = 2^m-1$ for which the generator is the minimal polynomial of a primitive element of $GF(2^m)$ is equivalent to the $(n,n-m)$-Hamming code.

Example: Using the irreducible polynomial $x^4 + x + 1$ we can generate GF(16). For the nonzero elements we find the representation:

1	α	α^2	α^3	α^4	α^5	α^6	α^7	α^8	α^9	α^{10}	α^{11}	α^{12}	α^{13}	α^{14}
1	0	0	0	1	0	0	1	1	0	1	0	1	1	1
0	1	0	0	1	1	0	1	0	1	1	1	1	0	0
0	0	1	0	0	1	1	0	1	0	1	1	1	1	0
0	0	0	1	0	0	1	1	0	1	0	1	1	1	1

If we use this representation of the (15,11) Hamming code and its parity-check matrix the encoder codes a sequence $(a_0,a_1,\ldots,a_{10})$ of information bits into the code word $c(x) := a(x)(x^4+x+1)$ where $a(x) := \sum_{i=0}^{10} a_ix^i$. Let us, as before, assume the received word contains one error, i.e. it is $c(x) + x^e$. The receiver computes the syndrome which is nothing else than $c(\alpha) + \alpha^e = \alpha^e$. Hence if the syndrome is α^e the decoding rule is to assume an error in position e. Note that decoding is just as simple as in our first description of Hamming codes but now the encoder has a simplicity we did not have before!

As a second example we let $m_1(x)$ be as above and consider $g(x) := (x+1)m_1(x)$. Then $\mathcal{R}g(x)$ is the cyclic code consisting of the words $c(x)$ for which $c(1) = c(\alpha) = 0$. This is a subcode of the Hamming code obtained by adding a row of 1's to the parity check matrix, i.e. it is the code consisting of the code words of even weight in the

Hamming code, (see (2.6.3)).

As a third example we take $n = 15$ and $g(x) = m_1(x)m_5(x)$. We have $g(x) = (x^4+x+1)(x-\alpha^5)(x-\alpha^{10}) = (x^4+x+1)(x^2+x+1) = 1 + x^3 + x^4 + x^5 + x^6$. Therefore $g(x)$ generates a 9-dimensional cyclic code of length 15. The simplest form of a parity-check matrix of this code now is

$$H = \begin{pmatrix} 1 & \alpha & \alpha^2 & \cdots\cdots & \alpha^{14} \\ 1 & \alpha^5 & \alpha^{10} & \cdots\cdots & \alpha^{70} \end{pmatrix}$$

where each entry α^i stands for a column vector with 4 entries from GF(2). Since $g(x)$ has degree 6 we know that a parity-check matrix for this code should have 6 rows. Therefore the rows of the matrix H are not independent.

Up to now we have taken $n = 2^m - 1$. Now let n be arbitrary, $q = 2$. Then $x^n - 1 = \prod_{i=1}^{n} (x - \beta^i)$ where β is a primitive n-th root of unity. If m is the multiplicative order of 2 mod n ($2^m \equiv 1 \bmod n$) and α is a primitive element of $GF(2^m)$, we can take $\beta = \alpha^{\frac{2^m-1}{n}j}$ where $(j,n) = 1$, i.e. β is any zero of the cyclotomic polynomial $Q^{(n)}(x)$. We can now specify a cyclic code by specifying β and giving the set $K := \{k (\mathrm{mod}\ n) \mid g(\beta^k) = 0\}$. Since $g(\beta^i) = 0 \Rightarrow g(\beta^{2i}) = 0$ we see that the permutation π_2 must leave K invariant (as a set). Vice versa any set K with this property defines a cyclic code (if β is given). We now show that in a sense, the code does not depend on the choice of β.

(3.2.2) <u>THEOREM</u>: <u>Let</u> β <u>and</u> β^* <u>be two primitive</u> n-th <u>roots of unity and let</u> K <u>be a subset of</u> $\{0,1,\ldots,n-1\}$ <u>which is closed under multiplication by</u> 2 (<u>mod</u> n). <u>Define</u> $g(x) := \prod_{k \in K} (x - \beta^k)$, $g^*(x) := \prod_{k \in K} (x - \beta^{*k})$. <u>Then</u> $g(x)$ <u>and</u> $g^*(x)$ <u>generate equivalent cyclic codes</u>.

<u>Proof</u>: There is a j with $(n,j) = 1$ such that $\beta^* = \beta^j$. If $c(x)$ is a code word in $\mathcal{R}g(x)$, say $c(x) = \sum_{i=0}^{n-1} c_i x^i$, then for every $k \in K$ we have $c(\beta^k) = 0$. This can be written as $\sum_{i=0}^{n-1} c_{\pi_j(i)} \beta^{ijk} = 0$, i.e. $\sum_{i=0}^{n-1} c_{\pi_j(i)} (\beta^{*k})^i = 0$.

This means that $(c_{\pi_j(0)}, c_{\pi_j(1)}, \ldots, c_{\pi_j(n-1)})$ is a code word in $\mathcal{R}g^*(x)$.

Therefore there is a permutation which permutes all code words of $\mathcal{R}g(x)$ into code words of $\mathcal{R}g^*(x)$.

The important thing we learn from the point of view of this paragraph is that the cyclic shifts are not the only permutations which leave a cyclic code invariant. If e.g. we take a binary cyclic code of odd word length n then the permutation π_2 permutes code words into code words. This information will be used to determine the weight enumerators for cyclic codes. (See 3.3). The larger the automorphism group of a code is the easier it is in general to find the weight enumerator of the code.

3.3 Idempotents

In this section we again use the symbol π_j to denote the permutation of $\{0,1,\ldots,n-1\}$ given by $\pi_j(k) = jk \pmod n$ and also to denote the automorphism of $\mathcal{R}$ given by $\pi_j(x^k) = x^{jk} \bmod (x^n-1)$. If V is the cyclic code $\mathcal{R}g(x)$ then $\pi_j V$ denotes the cyclic code $\mathcal{R}\pi_j(g(x))$. We shall now consider <u>binary</u> cyclic codes of length n (n odd). In Section 3.1 and 3.2 we saw that there is a 1-1 correspondence between the cycles of π_2 and the irreducible factors of $x^n - 1$. In fact if $(c_1, c_2, \ldots, c_k)$ is a cycle of π_2 and α is a primitive n-th root of unity then $f(x) = \prod_{i=1}^{k}(x - \alpha^{c_i})$ is an irreducible factor of $x^n - 1$.

(3.3.1) <u>THEOREM</u>: <u>For every ideal</u> V <u>in</u> $\mathcal{R}$ <u>there is a unique polynomial</u> $c(x) \in V$, <u>called the idempotent of</u> V, <u>with the following properties</u>:

(i) $c(x) = c^2(x)$,

(ii) $c(x)$ <u>generates</u> V,

(iii) $\forall_{f(x)\in V}\ [c(x)f(x) = f(x)]$, i.e. $c(x)$ <u>is a unit for</u> V,

(iv) if $(j,n) = 1$ <u>then</u> $\pi_j(c(x))$ <u>is the idempotent of</u> $\pi_j V$.

<u>Proof</u>: (i) Let $g(x)$ be the generator of V and $g(x)h(x) = x^n - 1$ (in R). Since $x^n - 1$ has no multiple zeros we have $(g(x),h(x)) = 1$. Therefore there are polynomials $p_1(x)$ and $p_2(x)$ such that (in R)

$$p_1(x)g(x) + p_2(x)h(x) = 1. \qquad (*)$$

Now take $c(x) := p_1(x)g(x)$. Multiply both sides of (*) by $c(x)$. We find $c^2(x) + p_1(x)p_2(x)g(x)h(x) = c(x)$. Since (in $\mathcal{R}$) $g(x)h(x) = 0$ we have proved (i).

(ii) The monic polynomial of lowest degree in the ideal generated by $c(x)$ is $(c(x),x^n-1) = (p_1(x)g(x),\ g(x)h(x)) = g(x)$.

(iii) By (ii) every $f(x) \in V$ is a multiple of $c(x)$. Let $f(x) = c(x)f_1(x)$. Then $c(x)f(x) = c^2(x)f_1(x) = c(x)f_1(x) = f(x)$, i.e. $c(x)$ is a unit in V.

(iv) Since π_j is an automorphism of $\mathcal{R}$ the polynomial $\pi_j(c(x))$ is an idempotent and unit in $\pi_j V$. Since the unit is unique we have proved (iv).

Note that an ideal can contain more than one idempotent but that only one of these has all the properties (i) to (iv). The advantage of using idempotents for describing cyclic codes is that it is no longer necessary to factor $x^n - 1$. To see this assume that $\sum_{i=1}^{k} x^{c_i}$ is an idempotent in $\mathcal{R}$. Then $\sum_{i=1}^{k} x^{2c_i} = \sum_{i=1}^{k} x^{c_i}$ which means that the set $\{c_1,c_2,\ldots,c_k\}$ is invariant under the permutation π_2 of $\{0,1,\ldots,n-1\}$. This set must then be the union of cycles of π_2. We have seen that the t cycles of π_2 are easily obtained and from these we find the 2^t idempotents generating the cyclic codes in $\mathcal{R}$.

Since we would like to use the representation of 3.1 and 3.2 and idempotents simultaneously we are faced with the problem of finding a correspondence between generators of ideals and the idempotents of the ideals. From the equation (*) we know that if $c(x)$ is the idempotent of V then $1 + c(x)$ is the idempotent of the dual code.

(3.3.2) DEFINITION: The idempotent of a minimal code M_i^- is called a <u>primitive idempotent</u> and denoted by $\theta_i(x)$ (or $\underline{\theta}_i$).

Remark: Notice that a minimal code is a finite field with the primitive idempotent as unit!

(3.3.3) <u>THEOREM</u>: <u>If V_1 and V_2 are ideals in $\mathcal{R}$ with idempotents $c_1(x)$ and $c_2(x)$ then</u>

(i) $V_1 \cap V_2$ <u>has idempotent</u> $c_1(x)c_2(x)$,

(ii) $V_1 + V_2$ <u>has idempotent</u> $c_1(x) + c_2(x) + c_1(x)c_2(x)$.

<u>Proof</u>: (i) $c_1(x)c_2(x)$ is a code word in $V_1 \cap V_2$. If $g_1(x)$ and $g_2(x)$ are generators of V_1 and V_2 then $V_1 \cap V_2$ is generated by their least common multiple $g(x)$. For any code word $a(x)g(x)$ we have $c_1(x)c_2(x)a(x)g(x) = c_2(x)a(x)g(x) = a(x)g(x)$, i.e. $c_1(x)c_2(x)$ is the (unique) unit of $V_1 \cap V_2$.

(ii) Obviously $\underline{c}_1 + \underline{c}_2 + \underline{c}_1\underline{c}_2$ is an idempotent in $V_1 + V_2$. Let $\underline{a}$ and $\underline{b}$ be arbitrary polynomials. Then

$$(\underline{c}_1 + \underline{c}_2 + \underline{c}_1\underline{c}_2)(\underline{a}\,\underline{c}_1 + \underline{b}\,\underline{c}_2) = \underline{a}\,\underline{c}_1 + \underline{b}\,\underline{c}_2\,,$$

i.e. $\underline{c}_1 + \underline{c}_2 + \underline{c}_1\underline{c}_2$ is the (unique) unit of $V_1 + V_2$.

(3.3.4) <u>THEOREM</u>: <u>For the primitive idempotents we have</u>:

(i) $\underline{\theta}_i\underline{\theta}_j = 0$ <u>if</u> $i \neq j$,

(ii) $\sum_{i=1}^{t} \underline{\theta}_i = 1$,

(iii) $1 + \underline{\theta}_{i_1} + \underline{\theta}_{i_2} + \cdots + \underline{\theta}_{i_r}$ <u>is the idempotent of the ideal</u> $\mathcal{R}f_{i_1}(x)f_{i_2}(x) \cdots f_{i_r}(x)$.

<u>Proof</u>: (i) By (3.3.2) and (3.3.3) $\underline{\theta}_i\underline{\theta}_j$ is the generator of $M_i^- \cap M_j^- = \{0\}$.

(ii) By repeated application of (3.3.3) (ii) and (3.3.4) (i) we see that $\sum_{i=1}^{t} \underline{\theta}_i$ is the idempotent of $M_1^- + M_2^- + \cdots + M_t^- = \mathcal{R}$, i.e. $\sum_{i=1}^{t} \underline{\theta}_i = 1$.

(iii) $\mathcal{R}f_{i_1}(x) \cdots f_{i_r}(x)$ is the dual of $M_{i_1}^- + M_{i_2}^- + \cdots + M_{i_r}^-$.

If S and T are nonempty subsets of $\{1,2,\ldots,t\}$ then $\left(\sum_{j\in S} \underline{\theta}_j\right)\left(\sum_{j\in T} \underline{\theta}_j\right) = \sum_{j\in S\cap T} \underline{\theta}_j$ and this is 0 only if $S \cap T = \emptyset$. Suppose $\underline{\xi}_1, \underline{\xi}_2, \ldots, \underline{\xi}_t$ is a set of idempotents for which $\underline{\xi}_i\,\underline{\xi}_j = 0$ if $i \neq j$. Every $\underline{\xi}_j$ is a sum of $\underline{\theta}_i$'s and no $\underline{\theta}_i$ can occur in two of the $\underline{\xi}_j$'s. Therefore the $\underline{\xi}_j$'s must be a permutation of the primitive idempotents.

This principle enables us to find the primitive idempotents in the following way:

(3.3.5) Algorithm:

First construct the idempotents $\eta_1, \eta_2, \ldots, \eta_t$ corresponding to cycles of π_2. Then $1 = \eta_1 + (1 + \eta_1)$ is a decomposition of 1 into mutually orthogonal idempotents. Suppose 1 has been written as the sum of $\tau < t$ mutually orthogonal idempotents, i.e. $1 = \sum_{j=1}^{\tau} \xi_j$ with $\xi_i\xi_j = 0$ if $i \neq j$. Let ξ be any idempotent. Suppose that for some j the idempotents $\xi_j\xi$ and $\xi_j(1+\xi)$ are different from 0. For $i \neq j$ we have $\xi_i\xi_j\xi = 0$ and $\xi_i\xi_j(1 + \xi) = 0$, i.e. the two idempotents $\xi_j\xi$ and $\xi_j(1 + \xi)$ are orthogonal to all the others and also mutually orthogonal and therefore all different. Since $\xi_j = \xi_j\xi + \xi_j(1 + \xi)$ we can then write 1 as the sum of $\tau + 1$ mutually orthogonal idempotents. This process fails if for every j we have $\xi_j\xi = 0$ or $\xi_j(1+\xi) = 0$ but since $\sum_{j=1}^{\tau} \xi_j = 1$ this means that ξ is a linear combination of the ξ_j's $(j = 1,\ldots,\tau)$. Among the idempotents $\eta_1, \eta_2, \ldots, \eta_t$ there is at least one for which this is not the case since the η_i form a basis for all the idempotents. This algorithm leads to a decomposition of 1 into t mutually orthogonal idempotents and these must then be the primitive idempotents.

(3.3.6) **Example**: Take n = 15. Let $\eta_1 = (1,2,4,8)$, $\eta_2 = (3,6,9,12)$, $\eta_3 = (5,10)$, $\eta_4 = (7,11,13,14)$, $\eta_5 = (0)$ where for abbreviation we have written (1,2,4,8) instead of $x + x^2 + x^4 + x^8$, etc. We start with $1 = \eta_1 + (1+\eta_1)$. Apply the algorithm with $\xi = \eta_2$ to $(1 + \eta_1)$. We find

$$1 = \eta_1 + (\eta_1 + \eta_2) + (1 + \eta_2).$$

Now apply the algorithm with η_3 to η_1 and $(1 + \eta_2)$. We find:

$$1 = (\eta_2+\eta_4) + (\eta_1+\eta_2+\eta_4) + (\eta_1+\eta_2) + (\eta_1+\eta_3+\eta_4) + (\eta_1+\eta_2+\eta_3+\eta_4+\eta_5)$$

which must be the decomposition into primitive idempotents. The last one of these was to be expected since for $g(x) = 1 + x$ the equation (*) is $(x^{n-2} + x^{n-4} + \cdots + x)(1 + x) + 1 \cdot (1 + x + \cdots + x^{n-1}) = 1$ and therefore $1 + x + \cdots + x^{n-1}$ is a primitive idempotent.

The advantage of the algorithm (3.3.5) is that it is easily carried out by computer.

We are now in a position to enumerate the weights of code words in a cyclic code. In the following T denotes the permutation $k \to k + 1 \pmod n$ or the mapping of $\mathcal{R}$ given by $T(g(x)) = xg(x)$. If $a(x)$ is a code word in a cyclic code then $T\underline{a}, T^2\underline{a}, \ldots, T^{n-1}\underline{a}$ are also code words in this code but not necessarily all different. The number of different code words in the set $\{\underline{a}, T\underline{a}, T^2\underline{a}, \ldots, T^{n-1}\underline{a}\}$ is called the <u>period</u> of $\underline{a}$ and denoted by per $(\underline{a})$.

(3.3.7) <u>LEMMA</u>: Let $a(x) \in \mathcal{R}$. In R let $(a(x), x^n-1) =: g(x)$, $h(x) := (x^n-1)/g(x)$. If e is the exponent of $h(x)$ then per $(\underline{a}) = e$.

<u>Proof</u>: (i) Define $g^*(x) := (x^e-1)/h(x)$. Let $a(x) = c(x)g(x)$ with $(c(x), x^n-1) = 1$. Then $a(x)(x^e-1) = c(x)g(x)g^*(x)h(x) = 0$ in $\mathcal{R}$, i.e. $T^e\underline{a} = \underline{a}$. Therefore per $(\underline{a}) \leq e$.

(ii) If per $(\underline{a}) = e'$ then $(x^{e'}-1)a(x) = 0$, i.e. $c(x)g(x)(x^{e'}-1)$ is divisible by x^n-1 in R. Since $(c(x), x^n-1) = 1$ we see that $h(x)$ divides $x^{e'} - 1$. Therefore $e' \geq e$.

From (i) and (ii) the lemma follows.

(3.3.8) <u>COROLLARY</u>: $\mathrm{per}(\underline{\theta}_i) = e_i$ where e_i is the exponent of $f_i(x)$.

The set $\{\underline{a}, T\underline{a}, \ldots, T^{e-1}\underline{a}\}$ is called a cycle of length e.

(3.3.9) <u>THEOREM</u>: <u>Every cycle (except the cycle containing $\underline{0}$) of M_i^- has length</u> e_i.

<u>Proof</u>: The theorem is a consequence of (3.3.7) since every nonzero code word in M_i^- and $x^n - 1$ have g.c.d. $(x^n-1)/f_i(x)$.

Now suppose we have cycle representatives for the codes M_i^- and M_j^-. E.g. let $a(x) \in M_i^-$, $b(x) \in M_j^-$, $h := (\mathrm{per}(\underline{a}), \mathrm{per}(\underline{b}))$, $H := [\mathrm{per}(\underline{a}), \mathrm{per}(\underline{b})]$. If $T^\alpha\underline{a} + T^\beta\underline{b} = T^{\alpha'}\underline{a} + T^{\beta'}\underline{b}$ then $T^\alpha\underline{a} - T^{\alpha'}\underline{a} = T^{\beta'}\underline{b} - T^\beta\underline{b}$, but since the word on the left-hand side is in M_i^- and the one on the right-hand side is in M_j^- they must both be $\underline{0}$. Therefore $T^\alpha\underline{a} + T^\beta\underline{b}$ takes on $\mathrm{per}(\underline{a})\cdot\mathrm{per}(\underline{b})$ different values. These are code words in $M_i^- + M_j^-$. Furthermore $\mathrm{per}(T^\alpha\underline{a} + T^\beta\underline{b}) = H$. Finally, suppose $T^\alpha\underline{a} + \underline{b}$ and $T^{\alpha'}\underline{a} + \underline{b}$ are in the same cycle of $M_i^- + M_j^-$. Then for some λ we have $T^\lambda(T^\alpha \underline{a} + \underline{b}) = T^{\alpha'} \underline{a} + \underline{b}$, i.e. $\lambda \equiv 0 \pmod h$ and therefore $\alpha \equiv \alpha' \pmod h$. Summarizing we have:

(3.3.10) THEOREM: If $\underline{a} \in M_i^-$ and $\underline{b} \in M_j^-$ then the per($\underline{a}$)·per($\underline{b}$) different code words $T^\alpha \underline{a} + T^\beta \underline{b}$ ($0 \le \alpha <$ per($\underline{a}$), $0 \le \beta <$ per ($\underline{b}$)) of the code $M_i^- + M_j^-$ are partitioned into h cycles of length H for which the words $T^\alpha \underline{a} + \underline{b}$ ($0 \le \alpha < h$) are representatives.

(3.3.11) Example: We shall now treat one example in detail to show that for codes with not too many code words, the weight enumerator can be determined by hand (or computer for slightly larger codes). Take n = 15. In (3.3.6) we found the five primitive idempotents to be

$$\underline{\theta}_1 := \underline{\eta}_2 + \underline{\eta}_4 = x^3+x^6+x^7+x^9+x^{11}+x^{12}+x^{13}+x^{14}$$
$$\underline{\theta}_2 := \underline{\eta}_1 + \underline{\eta}_2 + \underline{\eta}_4 = x+x^2+x^3+x^4+x^6+x^7+x^8+x^9+x^{11}+x^{12}+x^{13}+x^{14}$$
$$\underline{\theta}_3 := \underline{\eta}_1 + \underline{\eta}_2 = x+x^2+x^3+x^4+x^6+x^8+x^9+x^{12}$$
$$\underline{\theta}_4 := \underline{\eta}_1 + \underline{\eta}_3 + \underline{\eta}_4 = x+x^2+x^4+x^5+x^7+x^8+x^{10}+x^{11}+x^{13}+x^{14}$$
$$\underline{\theta}_5 := \underline{\eta}_1 + \underline{\eta}_2 + \underline{\eta}_3 + \underline{\eta}_4 + \underline{\eta}_5 = x+x^2+x^3+x^4+x^5+x^6+x^7+x^8+x^9+x^{10} + x^{11}+x^{12}+x^{13}+x^{14} .$$

We already know that $f_5(x) = 1 + x$. Since $(x^2 + x + 1)\theta_4(x) = 0$ we see that $f_4(x) = x^2 + x + 1$. We know from the cycles of π_2 that $x^{15} - 1 = (x+1)(x^2+x+1)\cdot Q^{(5)}(x)f_i(x)f_j(x)$ where $f_i(x)$ and $f_j(x)$ are the irreducible factors of $Q^{(15)}(x)$. Now $Q^{(5)}(x) = x^4 + x^3 + x^2 + x + 1$ corresponds to $\underline{\theta}_2$ i.e. $f_2(x) = 1 + x + x^2 + x^3 + x^4$. Consider the cyclic code $V := M_2^- + M_4^-$. Since f_2 has degree 4 and f_4 has degree 2 the code V has dimension 6. The idempotent of V is $\underline{\theta}_2 + \underline{\theta}_4 = \underline{\eta}_2 + \underline{\eta}_3 = x^3 + x^5 + x^6 + x^9 + x^{10} + x^{12} = x^3 \dfrac{x^{15} - 1}{(1+x+x^2)(1+x+x^2+x^3+x^4)}$. (In this case the idempotent is a cyclic shift of the generator!) The exponent of $f_2(x)$ is 5 and the exponent of $f_4(x)$ is 3. Therefore, by (3.3.9) the code M_2^- has three cycles of length 5 and the cycle consisting of $\underline{0}$ whereas the code M_4^- has one cycle of length 3 and the $\underline{0}$ cycle.

Notice that:

$\underline{\theta}_2 = (011110111101111)$,

$\underline{\theta}_2 + T\underline{\theta}_2 = (110001100011000)$,

$\underline{\theta}_2 + T^2\underline{\theta}_2 = (101001010010100)$

are obviously in different cycles of M_2^- and are therefore representatives of the three cycles of M_2^- of length 5. For M_4^- we have the representative $\underline{\theta}_4 = (011011011011011)$. We now apply Theorem (3.3.10) and find the following representatives of cycles for V:

$\underline{0} = (000000000000000)$,	weight	0 ,	length 1
$\underline{\theta}_4 = (011011011011011)$,	"	10 ,	" 3
$\underline{\theta}_2 = (011110111101111)$,	"	12 ,	" 5
$\underline{\theta}_2 + T\underline{\theta}_2 = (110001100011000)$,	"	6 ,	" 5
$\underline{\theta}_2 + T^2\underline{\theta}_2 = (101001010010100)$,	"	6 ,	" 5
$\underline{\theta}_4 + \underline{\theta}_2 = (000101100110100)$,	"	6 ,	" 15
$\underline{\theta}_4 + \theta_2 + T\underline{\theta}_2 = (101010111000011)$,	"	8 ,	" 15
$\underline{\theta}_4 + \underline{\theta}_2 + T^2\underline{\theta}_2 = (110010001001111)$,	"	8 ,	" 15 .

Therefore the weight enumerator of V is $1 + 25z^6 + 30z^8 + 3z^{10} + 5z^{12}$. The code is 2-error-correcting and 3-error-detecting.

Remark: The theory presented in this section is a special case of more general theorems on idempotents for an algebra with unity which is the supplementary sum of a number of ideals. We refer the interested reader to A. A. Albert, Structure of Algebras (Chapt. II), AMS Coll. Publ. 24. The special application in this section is due to F. J. MacWilliams (The structure and properties of binary cyclic alphabets, Bell System Tech. J. 44 (1965), 303-332).

3.4 Some other representations of cyclic codes

Let p be a prime and let k be the multiplicative order of p mod n. If $q = p^k$ then the primitive n-th roots of unity are in GF(q) and in no subfield of GF(q).

(3.4.1) DEFINITION: For $\xi \in GF(q)$ we define the <u>trace</u> of ξ as

$$Tr(\xi) := \xi + \xi^p + \xi^{p^2} + \cdots + \xi^{p^{k-1}} .$$

Note that the trace has the following properties:

(3.4.2) (a) For every $\xi \in GF(q)$ the trace $Tr(\xi)$ is in $GF(p)$.

(b) Since the equation $x + x^p + \cdots + x^{p^{k-1}} = 0$ has at most p^{k-1} roots in $GF(q)$ there are elements $\xi \in GF(q)$ with $Tr(\xi) \neq 0$.

(c) $\forall_{\xi \in GF(q)} \forall_{\eta \in GF(q)} [Tr(\xi + \eta) = Tr(\xi) + Tr(\eta)]$.

(d) $\forall_{a \in GF(p)} \forall_{\xi \in GF(q)} [Tr(a\xi) = a\, Tr(\xi)]$.

Combined with (c) this means that Tr is a linear mapping of the vector space $GF(q)$ over $GF(p)$ onto $GF(p)$.

(e) If $m_\xi(x)$ is the minimal polynomial of ξ and the degree of $m_\xi(x)$ is d then the polynomial $f_\xi(x) := \{m_\xi(x)\}^{k/d}$ is called the <u>field polynomial</u> of ξ. We remark that

$$f_\xi(x) = \prod_{i=0}^{k-1} (x - \xi^{p^i}) = x^k - Tr(\xi)x^{k-1} + \cdots + (-1)^k N(\xi).$$

In the regular representation of $GF(q)$ as a matrix ring over $GF(p)$ the trace of the matrix representing ξ is $Tr(\xi)$ and the characteristic polynomial of this matrix is $f_\xi(x)$.

Instead of Tr we sometimes shall use the notation $T_k(x) = x + x^p + x^{p^2} + \cdots + x^{p^{k-1}}$ which is useful if k is not fixed in the problem being discussed.

(3.4.3) <u>THEOREM</u>: <u>If β is a primitive</u> n-th <u>root of unity in</u> $GF(q)$ <u>where</u> $q = p^k$ (k <u>is the mult. order of</u> p <u>mod</u> n) <u>then the set</u>

$$V := \{\underline{c}(\xi) := (Tr(\xi), Tr(\xi\beta), \ldots, Tr(\xi\beta^{n-1})) \mid \xi \in GF(q)\}$$

<u>is an irreducible cyclic code</u> (<u>of length</u> n <u>and dimension</u> k).

__Proof:__ By (3.4.2) (c) and (d) V is a linear code. Next we note that $\underline{c}(\xi\beta^{-1})$ is a cyclic shift of $\underline{c}(\xi)$. Therefore V is a cyclic code. We know that β is a zero of an irreducible polynomial $h(x) = h_0 + h_1x + \cdots + h_kx^k$ since β is in no subfield of GF(q). If $\underline{c}(\xi) = (c_0, c_1, \ldots, c_{n-1}) \in V$ then $\sum_{i=0}^{k} c_ih_i = Tr(\xi h(\beta)) = Tr(0) = 0$, i.e. we have a parity check equation for the code V. Since h(x) is irreducible we see that $x^kh(x^{-1})$ is the check polynomial of V and V is therefore an irreducible code.

A __linear recurring sequence__ with elements in GF(q) is defined by an initial sequence $a_0, a_1, \ldots, a_{k-1}$ and a recursion

$$a_\ell + \sum_{i=1}^{k} b_ia_{\ell-i} = 0 \qquad (\ell \geq k). \tag{*}$$

The classical standard technique for finding a solution is to try $a_\ell = \beta^\ell$. This is a solution of (*) if β is a root of the equation

$$h(x) := x^k + \sum_{i=1}^{k} b_ix^{k-i} = 0.$$

Let us assume that this equation has k distinct roots $\beta_1, \beta_2, \ldots, \beta_k$ in some extension field of GF(q). Then, if $c_1, c_2, \ldots, c_k$ are arbitrary, the sequence $a_\ell = \sum_{i=1}^{k} c_i\beta_i^\ell$ is a solution of (*). We can then try to choose the c_i in such a way that $a_0, a_1, \ldots, a_{k-1}$ have the prescribed values. This means solving a system of k linear equations in the unknowns $a_0, a_1, \ldots, a_{k-1}$. The coefficient determinant is the Vandermonde determinant

$$\begin{vmatrix} 1 & 1 & \cdots & 1 \\ \beta_1 & \beta_2 & \cdots & \beta_k \\ \beta_1^2 & \beta_2^2 & \cdots & \beta_k^2 \\ \vdots & \vdots & & \vdots \\ \beta_1^{k-1} & \beta_2^{k-1} & \cdots & \beta_k^{k-1} \end{vmatrix} = \prod_{i>j} (\beta_i - \beta_j) \neq 0.$$

Hence we can solve the equations.

Now let h(x) be a divisor of $x^n - 1$, $(n,q) = 1$. The linear recurring sequence then is periodic with period n. If we consider the set of all $(a_0,a_1,\ldots,a_{n-1})$ where $(a_0,a_1,\ldots,a_{k-1})$ runs through all points of $\mathcal{R}^{(k)}$ this is an (n,k) cyclic code and by the definition we see that $x^k h(x^{-1})$ in the check polynomial. The presentation $a_\ell = \sum_{i=1}^{k} c_i\beta_i^\ell$ gives us another way of interpreting the code words.

We describe yet another representation of the code words in a cyclic code. Once again let $(n,q) = 1$ and let V be a cyclic code of length n over GF(q). Let β be a primitive n-th root of unity.

(3.4.4) DEFINITION: The <u>Mattson-Solomon</u> <u>polynomial</u> $MS(\underline{c},x)$ of a code word $\underline{c} \in V$ is defined as

$$MS(\underline{c},x) := \frac{1}{n} \sum_{j=1}^{n} S_j x^{n-j},$$

where $$S_j := c(\beta^j) = \sum_{i=0}^{n-1} c_i \beta^{ij} .$$

(We consider this polynomial as an element of $\mathcal{R} = R/S$, i.e. we have $x^n = 1$.) We then have

(3.4.5) <u>THEOREM</u>: <u>If</u> $\underline{c} = (c_0,c_1,\ldots,c_{n-1})$ <u>is a code word in a cyclic code</u> V <u>and the</u> S_j <u>are defined as in</u> (3.4.4) <u>then</u>

$$c_\ell = MS(\underline{c},\beta^\ell) \qquad (\ell = 0,1,\ldots,n-1).$$

<u>Proof</u>: $MS(\underline{c},\beta^\ell) = \frac{1}{n} \sum_{j=1}^{n} S_j\beta^{-j\ell} = \frac{1}{n} \sum_{j=1}^{n} \beta^{-j\ell} \sum_{i=0}^{n-1} c_i\beta^{ij} = \frac{1}{n} \sum_{i=0}^{n-1} c_i \sum_{j=1}^{n} \beta^{(i-\ell)j}.$

Since the inner sum is 0 if $i \neq \ell$ and n if $i = \ell$ the result follows.

To determine the degree of the Mattson-Solomon polynomial $MS(\underline{c},x)$ we consider the generator $g(x) = \prod_{k\in K} (x - \beta^k)$ of the code V. (Here $K \subset \{1,2,\ldots,n\}$ is invariant under π_q). If d is the smallest integer which is not in K then for every $j < d$ we have $c(\beta^j) = 0$, i.e. $S_j = 0$. This means that the degree of $MS(\underline{c},x)$ is $\leq n - d$. We have proved

(3.4.6) <u>LEMMA</u>: If V is a cyclic code generated by $g(x) = \prod_{k \in K} (x-\beta^k)$, if $\{1,2,\ldots,d-1\} \subset K$ and if $\underline{c} \in V$ then the degree of $MS(\underline{c},x)$ is at most $n - d$.

(3.4.7) <u>THEOREM</u>: <u>If there are</u> r <u>n-th roots of unity which are zeros of</u> $MS(\underline{c},x)$ <u>then</u> $w(\underline{c}) = n - r$.

<u>Proof</u>: The weight of $\underline{c}$ is the number of coordinates $c_\ell \neq 0$. By (3.4.5) there are r values of ℓ for which $c_\ell = 0$.

It is sometimes useful to rewrite the Mattson-Solomon polynomial in the following way. We consider only binary cyclic codes of length $n = 2^m - 1$. Then if α is a primitive element of $GF(2^m)$ and $\underline{c}$ a code word in a cyclic code V we have $S_j = c(\alpha^j)$. We now observe that $S_{2j} = S_j^2$. Now partition $\{1,\ldots,n\}$ into the cycles of π_2. Let deg (α^i) denote the degree of the minimal polynomial of α^i. Then we have (in $\mathcal{R}$):

(3.4.8)
$$MS(\underline{c},x) = \frac{1}{n} \sum_i{}^{*} T_{\deg(\alpha^i)}(S_i x^{-i})$$

where the * indicates that one i is taken from each cycle of π_2.

3.5 Problems

(3.5.1) Let V be a binary cyclic code with generator g(x). What condition must g(x) satisfy if changing all o's to 1's and vice versa leaves the code invariant?

(3.5.2) A (9,3) binary linear code V is defined by $(a_1,a_2,\ldots,a_9) \in V \Leftrightarrow (a_1 = a_2 = a_3 \text{ and } a_4 = a_5 = a_6 \text{ and } a_7 = a_8 = a_9)$. Show that V is equivalent to a cyclic code and determine the generator.

(3.5.3) A binary cyclic code of length 63 is generated by $x^5 + x^4 + 1$. What is the minimum distance of this code?

(3.5.4) A (63,47) binary cyclic code can be described by specifying some zeros of all code words as was shown in Section 3.2. How many zeros are necessary in this case?

(3.5.5) Consider the binary cyclic code of length 15 with check polynomial $x^6 + x^5 + x^4 + x^3 + 1$. Determine the weight enumerator of this code.

(3.5.6) Consider cyclic codes of length 11 over GF(3). Show that $f_1(x) := x - 1$, $f_2(x) := x^5 - x^3 + x^2 - x - 1$, $f_3(x) := x^5 + x^4 - x^3 + x^2 - 1$ are the irreducible factors of $x^{11} - 1$. Generalize section 3.3 to show that $-1 - x - x^2 \cdots -x^{10}$, $1 + x^2 + x^6 + x^7 + x^8 + x^{10}$ and $1 + x + x^3 + x^4 + x^5 + x^9$ are the primitive idempotents and that (3.3.4) (i) and (ii) hold.

Although the calculation is more lengthy it is possible to show, as in (3.3.11), that M_3^+ has weight enumerator $1 + 132z^5 + 132z^6 + 330z^8 + 110z^9 + 24z^{11}$. Prove that this code is perfect!

(3.5.7) Generate $GF(2^4)$ using the irreducible polynomial $x^4 + x + 1$. Let α be a primitive element of $GF(2^4)$. For $0 \leq i \leq 14$ write $\alpha^i = \sum_{j=0}^{3} a_{ij}\alpha^j$ with $a_{ij} \in GF(2)$ and let $\underline{r}_i$ denote the column vector $(a_{i0}, a_{i1}, a_{i2}, a_{i3})^T$. Let M_i be the matrix $(\underline{r}_i, \underline{r}_{i+1}, \underline{r}_{i+2}, \underline{r}_{i+3})$. Show that the 15 matrices M_i and the zero matrix with addition and matrix multiplication over GF(2) form a field isomorphic to $GF(2^4)$. For $0 \leq i \leq 14$ prove that $Tr(\alpha^i)$ = trace of $M_i = a_{i3}$.

IV. IMPORTANT CYCLIC CODES

4.1 BCH codes

We now introduce a set of multiple-error correcting codes which were discovered by R. C. Bose and D. K. Ray-Chaudhuri and independently by A. Hocquenghem and which are now known as BCH-codes.

Let $\mathcal{R} = \mathcal{R}^{(n)}$ as in Chapter 3 and let $g(x)$ be the generator of a cyclic code of length n over $GF(q)$. (Again we take $(n,q) = 1$). If m is the order of q mod n then let β be a primitive n-th root of unity in $GF(q^m)$.

(4.1.1) DEFINITION: If $g(x)$ is the product of the minimal polynomials of $\beta, \beta^2, \ldots, \beta^{d-1}$, (no factor occurring more than once), then the cyclic code generated by $g(x)$ is called a BCH code (of designed distance d).

We remark that a more general definition is used sometimes in which it is required that $g(x)$ is the product of the minimal polynomials of $\beta^{\ell}, \beta^{\ell+1}, \ldots, \beta^{\ell+d-2}$. If $n = q^m - 1$ we can take $\beta = \alpha$ where α is a primitive element of $GF(q^m)$. In this case the BCH code is called <u>primitive</u>.

We now show that the code defined in (4.1.1) has minimum distance at least d.

(4.1.2) <u>THEOREM</u>: <u>The minimum distance of a</u> BCH <u>code of designed distance</u> d <u>is at least</u> d.

1st <u>Proof</u>: We use the symbolic notation of Section 3.2 to define the $m(d-1)$ by n matrix H:

$$H := \begin{pmatrix} 1 & \beta & \beta^2 & \cdots & \beta^{n-1} \\ 1 & \beta^2 & \beta^4 & \cdots & \beta^{2(n-1)} \\ - & - & - & - & - \\ 1 & \beta^{d-1} & \beta^{2(d-1)} & \cdots & \beta^{(n-1)(d-1)} \end{pmatrix}$$

A word $\underline{c}$ is in the BCH code iff $\underline{c}H^T = \underline{o}$. Note that the $m(d-1)$ rows of H are not necessarily linearly independent. Consider any $d - 1$ columns of H.

The determinant of the submatrix of H obtained by taking the columns headed by $\xi_1 := \beta^{i_1}$, $\xi_2 := \beta^{i_2}$, ..., $\xi_{d-1} := \beta^{i_{d-1}}$ is

$$\begin{vmatrix} \xi_1 & \xi_2 & \cdots & \xi_{d-1} \\ \xi_1^2 & \xi_2^2 & \cdots & \xi_{d-1}^2 \\ - & - & - & - \\ \xi_1^{d-1} & \xi_2^{d-1} & \cdots & \xi_{d-1}^{d-1} \end{vmatrix} = \xi_1 \xi_2 \cdots \xi_{d-1} \prod_{i>j} (\xi_i - \xi_j) \neq 0$$

since β is a primitive n-th root of unity. Therefore any $d - 1$ columns of H are linearly independent and this implies that a code word $\underline{c} \neq \underline{o}$ has weight $w(\underline{c}) \geq d$.

2nd Proof: If $\underline{c}$ is a code word in the BCH-code then by Lemma (3.4.6) the degree of $MS(\underline{c},x)$ is at most $n - d$. Therefore in Theorem (3.4.7) $r \leq n - d$ and hence $w(\underline{c}) \geq d$.

The easiest case to treat is again the binary case. Note that since $g(\beta^i) = 0$ implies $g(\beta^{2i}) = 0$ every other row in H is superfluous. If $n = 2^m - 1$ and α is a primitive element of $GF(2^m)$ then a polynomial $g(x)$ for which $g(\alpha) = g(\alpha^3) = \cdots = g(\alpha^{2t-1}) = 0$ generates a cyclic code which by (4.1.2) corrects any pattern of up to t errors.

As an example we take $n = 31$ and $t = 4$. Now notice that $m_5(x) = (x-\alpha^5)(x-\alpha^{10})(x-\alpha^{20})(x-\alpha^9)(x-\alpha^{18}) = m_9(x)$. Therefore $g(x) = m_1(x)m_3(x)m_5(x)m_7(x)$ has as zeros a.o. $\alpha, \alpha^2, \alpha^3, \ldots, \alpha^{10}$ and hence $g(x)$ generates a 5-error-correcting code! This example shows that a BCH code can be better than (4.1.2) promises. Finding the actual minimal distance of a BCH code is in general a hard problem. Even the easier problem of finding which rows in the matrix H given above are superfluous is not easy. This problem amounts to finding the number of information symbols in the code. An algorithm which solves this problem was found by E. R. Berlekamp. It is given in [5], Chapter 12. The important thing to realize about BCH codes is that they have relatively high rates. Again, take $q = 2$, $n = 2^m - 1$.

If we demand $g(\alpha) = g(\alpha^3) = \cdots = g(\alpha^{2t-1}) = 0$ then $g(x)$ generates a t-error-correcting BCH code of length n. The parity-check matrix for this code can be obtained by taking a number of rows (maybe all the rows) of

$$H^* = \begin{pmatrix} 1 & \alpha & \alpha^2 & \cdots & \alpha^{n-1} \\ 1 & \alpha^3 & \alpha^6 & \cdots & \alpha^{3(n-1)} \\ 1 & \alpha^{2t-1} & & \cdots & \alpha^{(2t-1)(n-1)} \end{pmatrix}.$$

Therefore the rate of this code is $\geq 1 - \frac{mt}{2^m - 1}$.

Now suppose the rate R is really $1 - \frac{mt}{2^m - 1}$ and consider a BSC with error probability p. The expected number of errors per block at the receiver is pn and from what we learned in the last part of Section 2.2 we do not expect the BCH code to be very useful unless $t > pn$ which means $\frac{1 - R}{m} > p$. For a fixed rate R this is not true for large m. Therefore we do not expect a sequence of BCH codes to achieve what is promised by Shannon's theorem (1.4.3). Complete information on this problem is not known yet but the heuristic argument used above shows us what to expect. For practical purposes BCH codes are useful even though they are not the very good codes we are still looking for! We give one example of a more general BCH code: Let $n = 15$ and let α be a primitive element of $GF(2^4)$, (see the table on page 47). Then $\beta = \alpha^{14}$ is a primitive 15-th root of unity. If we require $g(\beta^0) = g(\beta^1) = g(\beta^2) = g(\beta^3) = g(\beta^4)$ then by an obvious extension of (4.1.2) $g(x)$ generates a BCH code with minimum distance ≥ 6. In this example $g(x) = m_0(x)m_3(x)m_7(x)$. The check polynomial is $m_1(x)m_5(x)$, i.e. $(x^4+x+1)(x^2+x+1) = x^6 + x^5 + x^4 + x^3 + 1$. Therefore this is the code of problem 3.5.5.

Let us now look at a procedure for correcting the errors. It is useful to take a simple example first. We take $n = 15$ and as before generate $GF(2^4)$ using $x^4 + x + 1$, (see the table on page 47). If α is a primitive element and $g(x) = m_1(x)m_3(x) = (x^4+x+1)(x^4+x^3+x^2+x+1) = x^8 + x^7 + x^6 + x^4 + 1$ then $g(x)$ generates the (15,7) binary 2-error-correcting BCH code. As a parity check matrix we can take

$$H^* := \begin{pmatrix} 1 & \alpha & \alpha^2 & \cdots & \alpha^{14} \\ 1 & \alpha^3 & \alpha^6 & \cdots & \alpha^{12} \end{pmatrix}.$$

Now let $\underline{x}$ be transmitted and $\underline{y} = \underline{x} + \underline{e}$ received. We assume that $\underline{e}$ has weight ≤ 2. Let e.g. the syndrome $\underline{y}H^{*T}$ be (11111101). Since this is not $\underline{0}$ we know at least one error has been made. If exactly one error was made say in the position corresponding to α^i then from the first row of H^* (i.e. actually the first 4 rows of H^* as (0,1)-matrix) we see that $\alpha^i = (1111)$ i.e. $i = 12$, Then $\alpha^{3i} = \alpha^{36} = \alpha^6 =$ (0011) and since this is not the second half of the syndrome the assumption that one error was made is false. So we turn to two errors, say in positions corresponding to α^i and α^j. Now $\underline{y}H^{*T} = (\underline{x}+\underline{e})H^{*T} = \underline{e}H^{*T}$ = (the transpose of) the sum of the columns i and j of H^*, i.e.

$$\alpha^i + \alpha^j = (1111) = \alpha^{12},$$
$$\alpha^{3i} + \alpha^{3j} = (1101) = \alpha^7 .$$

This gives us $\alpha^{i+j} = \{(\alpha^{3i}+\alpha^{3j}) + (\alpha^i+\alpha^j)^3\}/(\alpha^i+\alpha^j) = \alpha^{13}$. Therefore α^i and α^j are the solutions of the equation

$$\xi^2 + \alpha^{12}\xi + \alpha^{13} = 0.$$

We cannot solve this equation but by substituting $1, \alpha, \alpha^2, \ldots$ we find the roots α^3 and α^{10}. This means that two errors in positions corresponding to α^3 and α^{10} would indeed lead to the syndrome (11111101) and the decoder assumes this to be the error pattern. In practice the circuitry is designed in such a way that when the bit y_i is leaving the decoder this is changed to $y_i + 1$ iff α^i is a root of the equation. This simple example illustrates one of the difficulties of decoding. In our treatment the procedure depended on the number of errors and if we have e.g. a 5-error-correcting code this would become a complicated matter. The procedure described above, and also the one we describe below, will not always lead to decoding, i.e. failure can occur.

Now we shall describe a general decoding procedure. The presentation will be somewhat clearer if we use a different notation than we have been using. Consider a t-error-correcting BCH code. Let $\underline{C} = (C_0, C_1, \ldots, C_{n-1}) = C_0 + C_1x + \cdots + C_{n-1}x^{n-1}$ be a code word and $\underline{R} = R(x)$ the received word, $R(x) = C(x) + E(x)$ where $\underline{E} = (E_0, E_1, \ldots, E_{n-1})$ is the error-word. We use the matrix H as in (4.1.2). The syndrome then gives us $S_1, S_2, \ldots, S_{2t}$ where $S_k := R(\beta^k) = E(\beta^k) = \sum_{i=0}^{n-1} E_i\beta^{ik}$. If $e \leq t$ errors have occurred then e of the E_i are not zero. The positions where $E_i \neq 0$ are called the <u>error-locations</u>. If these correspond to $\beta^{j_1}, \beta^{j_2}, \ldots$ we shall write $X_1 := \beta^{j_1}, X_2 = \beta^{j_2}, \ldots$. The nonzero value of E_i at an error-location is called an error-value and these we denote by $Y_1, Y_2, \ldots, Y_e$.

(4.1.3) DEFINITION: The polynomial $\sigma(z) := \prod_{i=1}^{e}(1 - X_i z)$ is called the <u>error-locator</u>.

The zeros of $\sigma(z)$ are the reciprocals of the error-locations. Since S_k can be defined as was done above for all $k \geq 1$ we can introduce the formal power series

$$S(z) := \sum_{k=1}^{\infty} S_k z^k. \tag{4.1.4}$$

We then have:

$$S(z) = \sum_{k=1}^{\infty} S_k z^k = \sum_{k=1}^{\infty} z^k \sum_{i=1}^{e} Y_i X_i^k = \sum_{i=1}^{e} Y_i \sum_{i=1}^{\infty} (X_i z)^k = \sum_{i=1}^{e} \frac{Y_i X_i z}{1 - X_i z}$$

where all computations are with formal power series etc. Therefore we now have

$$S(z)\sigma(z) = \sum_{i=1}^{e} Y_i X_i z \prod_{j \neq i} (1 - X_j z) =: \omega(z), \tag{4.1.5}$$

where $\omega(z)$ is a polynomial of degree $\leq e$. We can rewrite (4.1.5) as a system of equations involving the known quantities $S_1, S_2, \ldots, S_{2t}$ and the unknowns $\sigma_0, \sigma_1, \ldots, \sigma_e$ if $\sigma(z) = \sum_{i=0}^{e} \sigma_i z^i$. Since the coefficients of $z^{e+1}, z^{e+2}, \ldots$ on the left-hand side of (4.1.5) must be 0 we find

$$\begin{aligned}
&\sigma_0 S_{e+1} + \sigma_1 S_e + \cdots + \sigma_e S_1 = 0\\
&\sigma_0 S_{e+2} + \sigma_1 S_{e+1} + \cdots + \sigma_e S_2 = 0\\
&-\,-\,-\,-\,-\,-\,-\,-\,-\,-\,-\,-\,-\,-\,-\,-\,-\\
&\sigma_0 S_{2t} + \sigma_1 S_{2t+1} + \cdots + \sigma_e S_{2t-e} = 0 \quad .
\end{aligned} \tag{4.1.6}$$

A decoding procedure then consists of solving (4.1.6) which gives the error-locations X_i and $\omega(z)$. By substituting $z = X_i^{-1}$ in $\omega(z)$ we can then find Y_i. This problem has no mathematical difficulty and only the implementation presents a problem. But we have not solved the major difficulty yet, namely: the decoder does not know e. Since we are using maximum likelihood decoding the following theorem solves this problem for us in theory.

(4.1.7) THEOREM: Let $\hat{e}$ be the smallest integer such that there exists a polynomial $\hat{\sigma}(z)$ of degree $\leq \hat{e}$ with $\hat{\sigma}(0) = 1$ and such that in the product $\hat{\sigma}(z)S(z)$ the coefficients of $z^{\hat{e}+1}, z^{\hat{e}+2}, \ldots, z^{2t}$ are 0. Then $\hat{e} = e$ and $\hat{\sigma}(z) = \sigma(z)$.

Proof: Obviously $\hat{e} \leq e$. Now we have for $k = \hat{e}+1, \ldots, 2t$

$$0 = \sum_{\ell=0}^{\hat{e}} \hat{\sigma}_\ell S_{k-\ell} = \sum_{\ell=0}^{\hat{e}} \hat{\sigma}_\ell \sum_{i=1}^{e} Y_i X_i^{k-\ell} =$$

$$= \sum_{i=1}^{e} Y_i X_i^k \sum_{\ell=0}^{\hat{e}} \hat{\sigma}_\ell X_i^{-\ell} =$$

$$= \sum_{i=1}^{e} \hat{\sigma}(X_i^{-1}) Y_i X_i^k .$$

We interpret this as a set of $2t - \hat{e}$ equations with coefficients X_i^k and unknowns $\hat{\sigma}(X_i^{-1})Y_i$. Since $2t - \hat{e} \geq 2t - e \geq e$ the columns of the coefficient matrix

$$\begin{pmatrix} X_1^{\hat{e}+1} & X_2^{\hat{e}+1} & \cdots & X_e^{\hat{e}+1} \\ X_1^{\hat{e}+2} & X_2^{\hat{e}+2} & \cdots & X_e^{\hat{e}+2} \\ - - - & - - - & - - - & - - - \\ X_1^{2t} & X_2^{2t} & \cdots & X_e^{2t} \end{pmatrix}$$

are independent (once again using the Vandermonde determinant!). Since the Y_i are not 0 all the $\hat{\sigma}(X_i^{-1})$ must be zero which proves the theorem.

Theorem (4.1.7) provides the key to the solution of the decoding problem. What we need is an algorithm which finds the polynomial $\hat{\sigma}$ of lowest degree satisfying the condition of (4.1.7) etc. Such an iterative algorithm was found by E. R. Berlekamp. The interested reader can find it in [5], Chapter 7 and a modification due to J. L. Massey in [6], Chapter 6. We shall not discuss it here nor go further into the problems of decoding BCH codes (failure etc).

In previous discussions on weight enumeration of codes we have remarked that to solve this problem it is useful to know the automorphism group of the code. In Section 3.3 we used the invariance of binary cyclic codes under the permutation π_2 to attack the problem of weight enumeration. For BCH codes we can find many other automorphisms as we shall now show. In the following we take $n = q^m - 1$ and we let α be a primitive element of $GF(q^m)$. We consider a BCH code of length n and extend it by adding an overall parity check. The overall parity-check symbol, first introduced in Section 2.2 is chosen such that in the extended word the sum of the symbols is 0. As before we shall denote the positions of the symbols in a word by X_i $(i = 0,1,\dots,n-1)$ where $X_i = \alpha^i$. We shall denote the additional position by X_∞ and we denote the element $0 \in GF(q^m)$ by α^∞. When representing a code word as a "polynomial" $C_0 + C_1x + \dots + C_{n-1}x^{n-1} + C_\infty x^\infty$ we shall let $1^\infty := 1$ and $(\alpha^i)^\infty = 0$ for $i \not\equiv 0 \pmod n$. The permutations $P_{u,v}$ of the elements of $GF(q^m)$ given by

$$P_{u,v}(X) := uX + v \quad (u \in GF(q^m),\ v \in GF(q^m),\ u \neq 0)$$

form a group, the <u>affine permutation group on</u> $GF(q^m)$. This group is doubly transitive (i.e. for every pair X_1, X_2 with $X_1 \neq X_2$ and every pair Y_1, Y_2 with $Y_1 \neq Y_2$ there is a $P_{u,v}$ such that $P_{u,v}(X_i) = Y_i$ $(i = 1,2)$). Notice that $P_{\alpha,0}$ is the cyclic shift on $X_0, X_1, \dots, X_{n-1}$ and that it leaves X_∞ invariant. Hence $P_{\alpha,0}$ transforms every extended cyclic code into itself. We now prove the following theorem due to W. W. Peterson.

(4.1.8) <u>THEOREM</u>: <u>Every extended</u> BCH <u>code of block length</u> $n + 1 = q^m$ <u>over</u> $GF(q)$ <u>is invariant under the affine permutation group on</u> $GF(q^m)$.

Proof: Let $\alpha, \alpha^2, \ldots, \alpha^{d-1}$ be the prescribed zeros of the generator of the BCH code. If $(C_0, C_1, \ldots, C_{n-1}, C_\infty)$ is a code word in the extended code and we apply the permutation $P_{u,v}$ to the positions of the symbols to obtain the permuted word $(C_0', C_1', \ldots)$ then for $k = 0, 1, \ldots, d-1$ we have

$$\sum_i C_i' \alpha^{ik} = \sum_i C_i (u\alpha^i + v)^k = \sum_i C_i \sum_{\ell=0}^{k} \binom{k}{\ell} u^\ell \alpha^{i\ell} v^{k-\ell} =$$

$$= \sum_{\ell=0}^{k} \binom{k}{\ell} u^\ell v^{k-\ell} \sum_i C_i (\alpha^\ell)^i = 0$$

because the inner sum $S_\ell := \sum_i C_i (\alpha^\ell)^i$ is 0 for $\ell = 0, 1, \ldots, d-1$. Therefore the permuted word is also in the extended code.

Observe that in the last line of the proof we would also get 0 if for those ℓ for which $S_\ell \neq 0$ we would have $\binom{k}{\ell} = 0$. This immediately yields the following generalization of (4.1.8) due to T. Kasami.

(4.1.9) THEOREM: Let α be a primitive element of $GF(q^m)$. If $g(x) = \prod_{k \in K} (x - \alpha^k)$ is the generator of a cyclic code of length $n = q^m - 1$ over $GF(q)$, if $0 \notin K$ and if $\forall_{k \in K} \forall_\ell \, [\binom{k}{\ell} \neq 0 \Rightarrow \ell \in K \cup \{0\}]$ then the extended code is invariant under the affine permutation group on $GF(q^m)$.

The permutation π_q which also leaves the extended code invariant is given by $\pi_q(X) = X^q$ and this permutation is not an element of the affine permutation group (unless $m = 1$ in which case π_q is the identity).

We shall now consider one example in detail. As usual we take $n = 15$ and generate $GF(2^4)$ using $x^4 + x + 1$ as in the table on page 47. Now consider the 2-error-correcting BCH code. This code is generated by $m_1(x)m_3(x) = (x^4+x+1)(x^4+x^3+x^2+x+1)$ which in the terminology of Section 3.3 is $f_3(x)f_2(x)$. The code is therefore $M_1^- + M_4^- + M_5^-$. Now apply the method of Section 3.3 to M_1^- and M_4^- and then observe that adding on M_5^- amounts to doubling the number of code words by adding on the all-1 vector to the generator matrix. We find the following set of representatives for the cycles of the BCH code and the connected words in the extended code:

Cycle representative $(C_0,C_1,\dots,C_{14})$	C_∞	length of cycle	weight	weight in extended code
$\underline{0}$ = (000000000000000)	0	1	0	0
$\underline{\theta}_5$ = (111111111111111)	1	1	15	16
$\underline{\theta}_4$ = (011011011011011)	0	3	10	10
$\underline{\theta}_4 + \underline{\theta}_5$ = (100100100100100)	1	3	5	6
$\underline{\theta}_1$ = (000100110101111)	0	15	8	8
$\underline{\theta}_1 + \underline{\theta}_5$ = (111011001010000)	1	15	7	8
$\underline{\theta}_1 + \underline{\theta}_4$ = (011111101110100)	0	15	10	10
$\underline{\theta}_1 + \underline{\theta}_4 + \underline{\theta}_5$ = (100000010001011)	1	15	5	6
$T\underline{\theta}_1 + \theta_4$ = (111001000001100)	0	15	6	6
$T\underline{\theta}_1 + \underline{\theta}_4 + \underline{\theta}_5$ = (000110111110011)	1	15	9	10
$T^2\underline{\theta}_1 + \theta_4$ = (101010010110000)	0	15	6	6
$T^2\underline{\theta}_1 + \underline{\theta}_4 + \underline{\theta}_5$ = (010101101001111)	1	15	9	10

The weight enumerator of the BCH code is $1 + 18z^5 + 30z^6 + 15z^7 + 15z^8 + 30z^9 + 18z^{10} + z^{15}$ and the weight enumerator for the extended code is $1 + 48z^6 + 30z^{15} + 48z^{10} + z^{16}$. The permutations $P_{u,v}$ are generated by $P_{\alpha,0}$ and the four "translations" $P_{1,1}$, $P_{1,\alpha}$, P_{1,α^2}, P_{1,α^3}. Using the table on page 47 we find

$P_{u,v}$	permutation of $(X_0, X_1, \dots, X_{14}, X_\infty)$
$P_{1,1}$	$(0,\infty)(1,4)(2,8)(3,14)(5,10)(6,13)(7,9)(11,12)$
$P_{1,\alpha}$	$(0,4)(1,\infty)(2,5)(3,9)(6,11)(7,14)(8,10)(12,13)$
P_{1,α^2}	$(0,8)(1,5)(2,\infty)(3,6)(4,10)(7,12)(9,11)(13,14)$
P_{1,α^3}	$(0,14)(1,9)(2,6)(3,\infty)(4,7)(5,11)(8,13)(10,12)$

If we examine the action of these permutations on $\underline{\theta}_1$ we see that the first three leave $(\underline{\theta}_1, 0)$ invariant and P_{1,α^3} transforms it into $(\underline{\theta}_1 + \underline{\theta}_5, 1)$. Since the two cycles $\underline{\theta}_1$ and $\underline{\theta}_1 + \underline{\theta}_5$ with $\underline{0}$ and $\underline{\theta}_5$ form the code words of $M_1^- + M_5^-$ which is the 3-error-correcting BCH code we could have expected this kind of behavior from (4.1.8). As a second example we apply $P_{1,\alpha}$ to $(T\underline{\theta}_1 + \theta_4, 0)$. This yields (0010111000000101) which is $(T^6(\underline{\theta}_1 + \underline{\theta}_4 + \underline{\theta}_5), 1)$.

The following theorem, due to E. Prange, shows how knowledge of the type of Theorem (4.1.8) can be used in weight enumeration.

(4.1.10) THEOREM: If $a(z) := \sum_{i=0}^{n} a_i z^i$ is the weight enumerator of a linear binary code of block length n and $A(z) := \sum_{i=0}^{n+1} A_i z^i$ is the weight enumerator of the extended code and if this extended code is invariant under a transitive permutation group then for all i

$$a_{2i-1} = \frac{2i\, a_{2i}}{n+1-2i} = \frac{2i}{n+1} A_{2i},$$

i.e. $$a(z) = A(z) + \frac{1-z}{n+1} A'(z).$$

Proof: Clearly $A_{2i} = a_{2i-1} + a_{2i}$ and $A_{2i-1} = 0$. In the extended code a_{2i-1} words of weight 2i have a 1 in position α^∞. The total weight of the code words of weight 2i is $2iA_{2i}$. If the extended code is invariant under a transitive permutation group this weight is divided equally among the n+1 positions, i.e. $(n+1)a_{2i-1} = 2iA_{2i}$ from which the theorem follows.

The weight enumerators of the 2-error-correcting BCH code which we calculated above are an example. Combining (4.1.8) and (4.1.10) we find

(4.1.11) THEOREM: The minimum weight of a binary BCH code is odd if $m = 2^m - 1$.

4.2 Reed-Solomon codes

Up to now we have been concerned mostly with random errors and we have paid no attention to the problem of correcting burst errors. We now consider codes introduced by I. S. Reed and G. Solomon which turned out to be a special kind of BCH codes and with which the problem of burst error correction can be attacked.

First we shall define what we mean by a burst error.

(4.2.1) DEFINITION: A burst of length d is a vector whose only nonzero components are among d successive components of which the first and the last are not zero. A burst error is an error vector which is a burst.

The following is a very simple theorem concerning these errors.

(4.2.2) THEOREM: Every (n,k) cyclic code can detect any burst error of length $d \leq n - k$.

Proof: An error vector is detected if it is not a code word. A burst of length $d \leq n - k$ has the form $x^i a(x)$ where $a(x)$ has degree $d - 1 \leq n - k - 1$. Hence $a(x)$ is not divisible by $g(x)$, the generator of the (n,k) code, since $g(x)$ has degree $n - k$. Therefore no burst vector of length $d \leq n - k$ is a code word.

We now first introduce the Reed-Solomon codes (RS-codes) and we will come back to the problem of burst errors later.

(4.2.3) DEFINITION: A Reed-Solomon code is a BCH code of length $n = q - 1$ over GF(q).

By (4.1.1) the generator of the code has the form $g(x) = \prod_{i=1}^{d-1} (x-\alpha^i)$ where α is a primitive element of GF(q) (which then is a primitive n-th root of unity). Clearly RS codes are only interesting for large q (the binary RS code has word length 1). The dimension of the code with generator of degree $d - 1$ is $k = n + 1 - d$. From (4.1.2) we know that the minimum distance of this RS code is at least d. Now let us consider any linear code of block length n and dimension k. If we consider $k - 1$ positions and look at the code words which have zeros in these positions we obtain a subcode of dimension at least 1 (since the $k - 1$ extra parity checks $a_{i_1} = a_{i_2} = \dots = a_{i_{k-1}} = 0$ are not necessarily independent of the other parity checks). This subcode has minimum distance at most $n - k + 1$ (namely the block length) and therefore this is also true for the original code. (An other way of seeing this is by remarking that every set of $n - k + 1$ columns of the parity check matrix $H = (-P^T I_{n-k})$ is linearly dependent. Hence $d \leq n - k + 1$.) From what we remarked above we now see that we have proved:

(4.2.4) **THEOREM**: *The minimum distance of the RS code with generator*

$$g(x) = \prod_{i=1}^{d-1} (x-\alpha^i) \text{ is } d.$$

The RS codes are examples of codes which are sometimes called *optimal codes*, i.e. (n,k) codes for which the minimum distance is $d = n - k + 1$ (which is the a priori maximum).

If we consider a set of d positions and then look at the subcode of an optimal code with zeros in all other positions this subcode has dimension $\geq k - (n-d) = 1$. Since this subcode has minimum distance d it must have dimension exactly 1. It follows that, for $n \geq d' \geq d$, specifying a set of d' positions and requiring the code words to be 0 in all other positions will define a subcode of dimension $d' - d + 1$ of the optimal code, i.e.

(4.2.5) **THEOREM**: *For* $n \geq d' \geq d = n + 1 - k$ *the subcode of an optimal* (n,k) *code consisting of those code words which have zeros outside a set of* d' *positions has dimension* $d' - d + 1$.

For a RS code the Mattson Solomon polynomial of a code word $\underline{c}$ has degree $\leq n - d = k - 1$ by (3.4.6). This polynomial has coefficients in $GF(q)$. This provides us with a way to encode information. The sequence $(a_0, a_1, \ldots, a_{k-1})$ of information bits is coded into $\underline{c} = (F(1), F(\alpha), \ldots, F(\alpha^{q-2}))$ where $F(x) = a_0 + a_1 x + \cdots + a_{k-1}x^{k-1} = MS(\underline{c},x)$. The RS codes are used in the following way. A t-error-correcting RS code over $GF(q^m)$ has word length $n = q^m - 1$ and dimension $n - 2t$. If we write each symbol in $GF(q^m)$ as a vector of length m over $GF(q)$ we obtain a new linear code of length nm and dimension $(n-2t)m$ over $GF(q)$. This code is then a good code to use to correct burst errors since a burst of length $b \leq (t-1)m + 1$ affects at most t consecutive symbols of the original code, an error which is corrected. Of course this code can also correct t random errors but that is not very good. On the other hand if several bursts occur which together affect $\leq$ t symbols of the original code they are all corrected! RS codes compare favorably with some other codes used for burst error correction.

It is possible to find the weight enumerator of an optimal code using (4.2.5) and the principle of inclusion and exclusion. To do this we first prove an analog of the Möbius inversion formula.

(4.2.6) DEFINITION: If N is a finite set and $S \subset T \subset N$ then we define

$$\mu(S,T) := (-1)^{|T|-|S|}.$$

(4.2.7) THEOREM: Let N be a finite set and let f be a function defined on P(N). If

$$g(S) := \sum_{R \subset S} f(R)$$

then

$$f(T) = \sum_{S \subset T} \mu(S,T) g(S).$$

Proof: $\sum_{S \subset T} \mu(S,T)g(S) = \sum_{S \subset T} \mu(S,T) \sum_{R \subset S} f(R) =$

$$= \sum_{R \subset T} f(R) \sum_{R \subset S \subset T} \mu(S,T)$$

and the proof now follows from the equality

$$\sum_{R \subset S \subset T} \mu(S,T) = \sum_{j=0}^{|T|-|R|} \binom{|T|-|R|}{j}(-1)^j = (1-1)^{|T|-|R|} = \begin{cases} 0 & \text{if } R \neq T, \\ 1 & \text{if } R = T. \end{cases}$$

Now we have

(4.2.8) THEOREM: Let V be an (n,k) optimal code over GF(q) and let $1 + \sum_{i=d}^{n} A_i z^i$, where $d = n + 1 - k$, be the weight enumerator of V. Then

$$A_i = \binom{n}{i} \sum_{j=0}^{i-d} (-1)^j \binom{i}{j}(q^{i-j-d+1} - 1)$$

$$= \binom{n}{i}(q-1) \sum_{j=0}^{i-d} (-1)^j \binom{i-1}{j} q^{i-j-d} \qquad (i = d,d+1,\ldots,n).$$

Proof: Take $N := \{0,1,\ldots,n-1\}$ and for $R \in P(N)$ define $f(R)$:= number of code words $(c_0,c_1,\ldots,c_{n-1})$ for which $c_i \neq 0 \Leftrightarrow i \in R$. If we define g as in (4.2.7) we have by (4.2.5)

$$g(S) = \begin{cases} 1 & \text{if } |S| \leq d-1, \\ q^{|S|-d+1} & \text{if } n \geq |S| \geq d . \end{cases}$$

By our definition of f we have $A_i = \sum_{R \subset N, |R|=i} f(R)$ and therefore application of (4.2.7) yields

$$A_i = \sum_{R \subset N, |R|=i} \sum_{S \subset R} (-1)^{|R|-|S|} g(S) =$$

$$= \binom{n}{i}\left\{\sum_{j=0}^{d-1} \binom{i}{j}(-1)^{i-j} + \sum_{j=d}^{i} \binom{i}{j}(-1)^{i-j} q^{j-d+1}\right\} =$$

$$= \binom{n}{i}\left\{\sum_{j=d}^{i} \binom{i}{j}(-1)^{i-j}\left(q^{j-d+1} - 1\right)\right\} = \binom{n}{i} \sum_{j=0}^{i-d} \binom{i}{j}(-1)^j\left(q^{i-j-d+1} - 1\right)$$

from which the second assertion of the theorem follows if we replace $\binom{i}{j}$ by $\binom{i-1}{j-1} + \binom{i-1}{j}$ and take together the terms with $\binom{i}{j}$.

(4.2.9) COROLLARY: If an optimal (n,k) code over GF(q) exists then $q \geq n - k + 1$ or $k \leq 1$.

Proof: Let $d = n - k + 1$. By (4.2.8) we have (if $d < n$) $0 \leq A_{d+1} = \binom{n}{d+1}(q-1)(q-d)$.

Another inequality of this kind can be obtained by considering a set $R \subset N$ with $|R| = d - 2$. Fix a position outside R. By (4.2.5) we can find a code word which has a 1 in this position, nonzero coefficients in the positions of R and one other nonzero coefficient in any of the remaining positions. Since two such code words have distance $\geq d$ the parts corresponding to the positions of R must have distance $d - 2$ for any pair of these words. This is impossible if there are more than $q - 1$ of these words. This proves:

(4.2.10) THEOREM: If an optimal (n,k) code over $GF(q)$ exists then $q \geq k + 1$ or

$$d = n - k + 1 \leq 2.$$

Actually (4.2.9) and (4.2.10) are dual theorems. This is shown by

(4.2.11) THEOREM: If an (n,k) code is optimal then the dual code is also optimal.

> Proof: Let $G = (I_k P)$ be the generator matrix of an (n,k) code. This code is optimal iff every set of $d - 1 = n - k$ columns of the parity-check matrix $H = (-P^T\ I_{n-k})$ is linearly independent and this is equivalent to saying that every square submatrix of $-P^T$ has a nonzero determinant. This then also holds for P and since G is the parity check matrix of the dual code this code is also optimal.

4.3 Generalized Reed-Muller codes

We now consider a class of (extended) cyclic codes some of which will turn out to be equivalent to the RM codes defined in (2.3.6). First, we generalize the idea of Hamming-weight to integers written in the q-ary number system:

(4.3.1) DEFINITION: If q is an integer ≥ 2 and $j = \sum_{i=0}^{m-1} \xi_i q^i$, with $0 \leq \xi_i < q$ for

$$i = 0, 1, \ldots, m-1, \text{ then we define } w_q(j) := \sum_{i=0}^{m-1} \xi_i.$$

Remark: Note that the sum in (4.3.1) is taken in $\mathbb{Z}$.

The codes we wish to study are now defined as follows:

(4.3.2) DEFINITION: The shortened r-th order generalized Reed-Muller code (GRM code) of length $n = q^m - 1$ over $GF(q)$ is the cyclic code with generator

$$g(x) := \prod_{\substack{0<j<q^m-1 \\ 0\leq w_q(j)<(q-1)m-r}} (x - \alpha^j),$$

where α is a primitive element in $GF(q^m)$. The <u>r-th order GRM code</u> of length q^m has the generator matrix

$$G^* := \begin{pmatrix} 1 & 1 \; 1 \; 1 \; \cdots \; 1 \\ 0 & \\ 0 & \\ \vdots & G \\ 0 & \end{pmatrix},$$

where G is a generator matrix for the shortened GRM code.

(It is obvious that the set of exponents j in (4.3.2) is closed under multiplication by q). Notice that by (3.2.2) replacing α by α^{-1} in (4.3.2) yields an equivalent code. In terms of α this code has generator

$$g^*(x) := \prod_{\substack{0<j<q^m-1, \\ w_q(j)>r}} (x - \alpha^j)$$

and the dual code then has generator

$$h^*(x) := \prod_{\substack{0<j<q^m-1, \\ w_q(j)\leq r}} (x - \alpha^j).$$

From this we prove:

(4.3.3) <u>THEOREM</u>: <u>The dual of the</u> r-<u>th order GRM code of length</u> q^m <u>is equivalent to a</u> ((q-1)m-r-1)-st <u>order GRM code</u>.

<u>Proof</u>: From what we showed above we know that $(x-1)h^*(x)$ is the generator of a ((q-1)m-r-1)-st order shortened GRM code. If we now lengthen the cyclic codes to obtain the two GRM codes we find as generator matrices

$$\begin{pmatrix} 1 & 1 & \cdots & 1 \\ 0 & & & \\ \vdots & & G & \\ 0 & & & \end{pmatrix} \quad \text{and} \quad \begin{pmatrix} 1 & \cdots & 1 \\ 0 & & \\ \vdots & \hat{H} & \\ 0 & & \end{pmatrix}$$

where $\hat{H}$ is the generator matrix of the code with $(x-1)h^*(x)$ as generator polynomial. For the two matrices orthogonality of all rows except the first is a consequence of what we proved above. We now have to show that the all-1 word is orthogonal to itself. This is true since the word length is a multiple of q. For the other rows in the two matrices this orthogonality is a consequence of the factor $(x-1)$ in both of the generators. The proof is complete since the dimensions of the lengthened cyclic codes have sum q^m.

To show the connection of these codes with the codes defined in Section 2.3 we must first prove a lemma.

(4.3.4) LEMMA: Let $(n,q) = 1$. Let V_1 be a cyclic code of length n over GF(q) with check polynomial $f_1(x) = \prod_{i=1}^{k_1} (x-\alpha_i)$. Let V_2 be a cyclic code of the same length over GF(q) with check polynomial $f_2(x) = \prod_{j=1}^{k_2} (x-\beta_j)$. By $f_1(x) \otimes f_2(x)$ we denote the polynomial which divides $x^n - 1$ and which has as its zeros all products $\alpha_i\beta_j$ ($i = 1,2,\ldots,k_1$; $j = 1,2,\ldots, k_2$); (note that these products need not all be different). The set of words

$$\left\{(a_0b_0, a_1b_1, \ldots, a_{n-1}b_{n-1}) \,\middle|\, (a_0, a_1, \ldots, a_{n-1}) \in V_1,\ (b_0, \ldots, b_{n-1}) \in V_2\right\}$$

generates a cyclic code which is a subcode of the code with $f_1(x) \otimes f_2(x)$ as check polynomial.

<u>Proof</u>: The assertion is an immediate consequence of one of the representations of cyclic codes we studied in Section 3.4 namely the one using linear recurring sequences. We know there are constants $c_1, c_2, \ldots, c_{k_1}$ and $c_1', c_2', \ldots, c_{k_2}'$ such that for $\ell = 0, 1, \ldots, n-1$ we have

$$a_\ell = \sum_{i=1}^{k_1} c_i \alpha_i^{-\ell} \quad \text{and} \quad b_\ell = \sum_{j=1}^{k_2} c'_j \beta_j^{-\ell}$$

from which we find that for $\ell = 0, 1, \ldots, n-1$ the components $a_\ell b_\ell$ are linear combinations (with coefficients independent of ℓ) of the reciprocals of zeros of $f_1(x) \otimes f_2(x)$, proving the lemma.

Now the correspondence we referred to is:

(4.3.5) THEOREM: The r-th order GRM code of length 2^m over GF(2) is equivalent to the r-th order RM code of length 2^m.

Proof: The proof is by induction. Since for $r = 0$ the codes defined by (2.3.6) and (4.3.2) are repetition codes, they are identical. For $r = 1$ we have already proved (4.3.5) namely by a combination of (2.3.7), (2.3.8) and (3.2.1) and (4.3.3). Now assume (4.3.5) is true for some r. We can choose a primitive element $\alpha \in GF(2^m)$ in such a way that the check polynomial of the shortened r-th order GRM code is $h^*(x) = \prod_{\substack{0<j<2^m-1, \\ w_2(j) \leq r}} (x - \alpha^j)$. The check polynomial of the shortened 1-st order RM code is $h_1^*(x) := \prod_{\substack{0<j<2^m-1, \\ w_2(j) = 1}} (x - \alpha^j)$.

Now $h^*(x) \otimes h_1^*(x) = \prod_{\substack{0<j<2^m-1 \\ w_2(j) \leq r+1}} (x-\alpha^j)$, (for $r < m-2$). The procedure described in (4.3.4) corresponds to the way one goes from the r-th order RM code to the (r+1)-st order RM code by definition (2.3.6). Therefore (4.3.5) follows from (4.3.4).

In Sections 2.3 and 2.4 two error-correction procedures for RM codes were described. The method of threshold decoding can be used with a fixed scheme for all the positions of the shortened code if this one is represented in the manner of this section, making it a cyclic code. The decoding procedure described partially in Section 4.1 can also be used as we shall now show.

(4.3.6) THEOREM: The r-th order GRM code of length q^m over GF(q) is a subcode of the extended BCH code of designed distance $d = (q-R)q^{m-Q-1}$ where $r = Q(q-1) + R$, $0 \leq R < q-1$.

Proof: For all $j < d-1$ the number $w_q(j)$ is less than

$$w_q((q-R)q^{m-Q-1} - 1) = (q-R-1) + \sum_{i=0}^{m-Q-2} (q-1) =$$

$$= (q-1)(m-Q-1) + (q-R-1)$$

$$= (q-1)m - r.$$

Hence $\alpha^0, \alpha^1, \ldots, \alpha^{d-2}$ are among the zeros of the generator of the r-th order GRM code. (Here we use the more general definition of BCH code).

Note that (4.3.6) generalizes (2.3.14).

4.4 Quadratic residue codes

(4.4.1) Notation: (i) In this section n will denote an odd prime.

(ii) α is a primitive n-th root of unity in an extension field of GF(q).

(iii) R_0 denotes the set of quadratic residues mod n, i.e. those elements $\neq 0$ in GF(n) which are squares; R_1 is the set of nonzero elements of GF(n) which are not in R_0.

(iv) $g_0(x) := \prod_{r \in R_0} (x - \alpha^r)$, $g_1(x) := \prod_{r \in R_1} (x - \alpha^r)$.

Note that $x^n - 1 = (x-1)g_0(x)g_1(x)$.

In the following we shall assume that $q^{\frac{n-1}{2}} \equiv 1 \pmod{n}$ or in other words that q is a quadratic residue mod n.

The nonzero elements of GF(n) are powers of a primitive element $a \in GF(n)$. Clearly a^e is a quadratic residue if $e \equiv 0 \pmod 2$ and a non-residue if $e \equiv 1 \pmod 2$. The product of two (non-) residues is a quadratic residue. Since q is a quadratic residue the set R_0 is closed under multiplication by q (mod n).

We then are able to define:

(4.4.2) DEFINITION: The cyclic codes of length n over GF(q) with generators $g_0(x)$ and $(x-1)g_0(x)$ are both called quadratic residue codes (QR-codes). The extended quadratic residue code of length n + 1 is obtained by annexing an overall parity check to the code with generator $g_0(x)$.

We remark that in the binary case the code with generator $(x-1)g_0(x)$ consists of the words of even weight in the code with generator $g_0(x)$. If G is a generator matrix for the first of these codes then adding the all-one vector will yield a generator matrix for the second code and the generator matrix for the extended code will then be

$$\begin{pmatrix} 1 & 1 & 1 & \cdots & 1 \\ 0 & & & & \\ \vdots & & G & & \\ 0 & & & & \end{pmatrix} .$$

In the binary case the condition that q is a quadratic residue mod n is satisfied if $n \equiv \pm 1 \pmod 8$. To show this we consider the integers $1, 2, \ldots, \frac{n-1}{2}$. If we multiply all these by 2 we obtain the even integers $\leq \frac{n-1}{2}$ and the even integers $x > \frac{n-1}{2}$ which we can write as $x = -y$ in GF(n) where $1 \leq y \leq \frac{n-1}{2}$. The number of these is $\nu = \frac{n-1}{4}$ if $n \equiv 1 \pmod 4$ and $\nu = \frac{n+1}{4}$ if $n \equiv -1 \pmod 4$. Hence

$$2^{\frac{n-1}{2}} \prod_{i=1}^{(n-1)/2} i \equiv (-1)^{\nu} \prod_{j=1}^{(n-1)/2} j \pmod n$$

(it is easily seen that after the above reduction each factor $\leq \frac{n-1}{2}$ occurs once). Since the two products are not divisible by n we find that

$2^{\frac{n-1}{2}} \equiv 1 \pmod n$ if ν is even, i.e. $n \equiv \pm 1 \pmod 8$.

We saw above that if $j \in R_1$ the permutation π_j maps R_0 into R_1 and vice versa. From (3.2.2) it then follows that if we replace R_0 by R_1 in (4.4.2) we obtain equivalent codes. If $n \equiv -1 \pmod 4$ then $-1 \in R_1$ and therefore the transformation $x \to x^{-1}$ maps a code word of the code with generator $g_0(x)$ into a code word of the code with generator $g_1(x)$.

We now prove some theorems which will enable us to find inequalities for the minimum weight of QR-codes. Let us consider a code word $c(x) = \sum_{i=0}^{n-1} c_i x^i$ of the code with generator $g_0(x)$ for which $\sum_{i=0}^{n-1} c_i \neq 0$ (i.e. the code word does not belong to the QR-code with generator $(x-1)g_0(x)$). As we saw above a suitable π_j will permute $c(x)$ into a word $\hat{c}(x)$ which is divisible by $g_1(x)$ but again not by $(x-1)$. If the weight of $\underline{c}$ is d then the weight of $\hat{\underline{c}}$ is also d and the product $c(x)\hat{c}(x)$ is a polynomial with at most d^2 nonzero coefficients. Since $c(x)\hat{c}(x)$ is a multiple of $g_0(x)g_1(x)$ and not a multiple of $(x-1)$ it must be a constant multiple of $1 + x + x^2 + \cdots + x^{n-1}$. This proves the following theorem:

(4.4.3) THEOREM: If a code word $c(x) = \sum_{i=0}^{n-1} c_i x^i$ of the code with generator $g_0(x)$, for which $\sum_{i=0}^{n-1} c_i \neq 0$, has weight d then

$$d^2 \geq n.$$

Remark: Since n is prime, equality in (4.4.3) is impossible.

The following improvement is possible if $n \equiv -1 \pmod 4$. We remarked above that in this case we could take $\hat{c}(x) = c(x^{-1})$. In this case the product $c(x)\hat{c}(x)$ has d terms of the form $c_i^2 x^0$, i.e. $c(x)\hat{c}(x)$ has at most $d^2 - d + 1$ nonzero coefficients:

(4.4.4) THEOREM: Let $n \equiv -1 \pmod 4$. If a code word $c(x) = \sum_{i=0}^{n-1} c_i x^i$ of the QR-code with generator $g_0(x)$, for which $\sum_{i=0}^{n-1} c_i \neq 0$, has weight d then

$$d^2 - d + 1 \geq n.$$

This type of argument can be extended further. Suppose $c(x) = \sum_{i=1}^{d} x^{e_i}$ is a word of odd weight d in the binary QR-code with generator $g_0(x)$. Let $n \equiv -1 \pmod 8$. The product $c(x)\hat{c}(x) = 1 + \sum\sum_{i \neq j} x^{e_i - e_j}$ will have less than $d^2 - d + 1$ terms if some of the terms cancel, i.e. if $e_i - e_j = e_k - e_\ell$ for some i,j,k,ℓ. In that case also $e_j - e_i = e_\ell - e_k$. Apparently the number of terms which cancel is a multiple of 4. Therefore $d^2 - d + 1 - 4\ell = n$ which implies that $d \equiv 3 \pmod 4$:

(4.4.5) THEOREM: If $n \equiv -1 \pmod 8$ and $\underline{c}$ is a code word of odd weight d in the binary QR-code with generator $g_0(x)$ then $d \equiv 3 \pmod 4$.

We shall now study in more detail the binary QR-codes. Again n is a prime $\equiv \pm 1 \pmod 8$ and α is a primitive n-th root of unity in an extension field of GF(2). Let us consider the polynomial

$$\theta(x) := \sum_{r \in R_0} x^r .$$

Since R_0 is closed under multiplication by 2 we see that $\theta(x)$ is an idempotent in the ring $\mathcal{R}$ of polynomials over GF(2) mod $x^n - 1$. Hence $\{\theta(\alpha)\}^2 = \theta(\alpha)$ which implies that $\theta(\alpha) \in GF(2)$. In the same way we see that $\theta(\alpha^i) = \theta(\alpha)$ if $i \in R_0$. If, on the other hand, $i \in R_1$ then $\theta(\alpha^i) + \theta(\alpha) = \sum_{r \in R_0 \cup R_1} \alpha^r = \alpha + \alpha^2 + \cdots + \alpha^{n-1} = 1$. Since $\theta(\alpha)$ cannot be equal to one for all the primitive n-th roots of unity we can choose α in such a way that $\theta(\alpha) = 0$. This then implies that $\theta(\alpha^i) = 0$ if $i \in R_0$ and $\theta(\alpha^i) = 1$ if $i \in R_1$. If $i = 0$ then $\theta(\alpha^i) = \frac{n-1}{2}$ which is 0 if $n \equiv 1 \pmod 8$ and 1 if $n \equiv -1 \pmod 8$. It follows that $\theta(x)$ is a multiple of $(x-1)g_0(x)$ if $n \equiv 1 \pmod 8$ and that $\theta(x)$ is a multiple of $g_0(x)$ if $n \equiv -1 \pmod 8$ and in fact we have shown:

(4.4.6) THEOREM: If the primitive n-th root of unity α in (4.4.1) (ii) is suitably chosen then the polynomial $\theta(x) = \sum_{r \in R_0} x^r$ is the idempotent of the binary QR-code with generator $(x-1)g_0(x)$ if $n \equiv 1 \pmod 8$ and of the code with generator $g_0(x)$ if $n \equiv -1 \pmod 8$.

To obtain the extended code in the case $n \equiv 1 \pmod 8$ we must annex the all one vector to the generator matrix of the code with $(x-1)g_0(x)$ as generator. In the case $n \equiv -1 \pmod 8$ we just annex an overall parity check to the words of the QR-code with $g_0(x)$ as generator (see the remark following (4.4.2)). Now let C be the circulant matrix with the code word $\underset{\sim}{\theta}$ as its first row. Let $\underline{c} := (0\ 0\ \cdots\ 0)$ if $n \equiv 1 \pmod 8$ and $\underline{c} := (1\ 1\ 1\ \cdots\ 1)$ if $n \equiv -1 \pmod 8$ and define

$$G := \begin{pmatrix} 1 & 1 & 1 & \cdots & 1 \\ \underline{c}^T & & C & & \end{pmatrix} .$$

We then have the following consequence of (4.4.6):

(4.4.7) COROLLARY: The rows of G generate the extended binary QR-code of length $n + 1$.

Of course the rows of G are not linearly independent since there are twice as many as the dimension of the code. The representation of (4.4.7) will be useful in the following discussion on permutations which leave the extended QR-code invariant.

We shall number the positions of the symbols in the extended QR-code using the coordinates of the projective line of order n, i.e. ∞, 0, 1, ..., n-1. We make the usual conventions about arithmetic operations with these coordinates e.g. $\pm 0^{-1} = \infty$, $\pm\infty^{-1} = 0$, $\infty + a = \infty$ for $a = 0, 1, \ldots, n-1$ etc.

(4.4.8) <u>THEOREM</u>: <u>The extended binary QR-code of length</u> $n + 1$ <u>is invariant under the group</u> PSL (2,n) <u>acting on the positions</u>.

<u>Proof</u>: The group PSL(2,n) consists of all the transformations $x \to \frac{ax + b}{cx + d}$ with a,b,c,d in GF(n) and $ad - bc = 1$. This group is generated by the two transformations $Sx := x + 1$ and $Tx := -x^{-1}$. Since the transformation S leaves position ∞ invariant and since it is a cyclic shift of the other positions it leaves the code invariant. To study the action of T we use (4.4.7). Apply T to the rows and columns of G to obtain the matrix $H := [h_{ij}]$ which generates the permuted code. We consider a number of cases separately:

(i) If $i = 0$ then $h_{ij} = g_{\infty,-j^{-1}} = 1$ for all j, i.e. row 0 of H is equal to row ∞ of G.

(ii) If $i = \infty$ then $h_{ij} = g_{0,-j^{-1}}$. If $n \equiv 1 \pmod 8$ then $j \in R_0$ iff $-j^{-1} \in R_0$ since in this case -1 is a square and therefore $g_{0,0} = g_{0,\infty} = 0$ which proves that row ∞ of H is equal to row 0 of G. If, however, $n \equiv -1 \pmod 8$ then $j \in R_0$ iff $-j^{-1} \in R_1$ and $g_{0,0} = 0$ while $g_{0,\infty} = 1$. In that case $g_{0,-j^{-1}} = g_{\infty,j} + g_{0,j}$, i.e. row 0 of H is the sum of two rows of G.

(iii) Consider row i of H where $i \in R_0$. Add row i of G to this row. In the sum row we have zeros for $j = i$, for $j = \infty$ if $n \equiv -1 \pmod 8$ and for those $j \neq 0$ for which $j - i$ and $-j^{-1} + i^{-1}$ are both in R_0 or both in R_1. This last condition is satisfied iff $(j-i)(-j^{-1}+i^{-1}) \in R_0$, i.e. $ij \in R_0$, i.e. $j \in R_0$. This proves that the sum row is equal to the sum of rows 0 and ∞ of G, i.e. row i of H is the sum of 3 rows of G.

(iv) We leave as an exercise to the reader to show that if $i \in R_1$ then row i of H is the sum of 2 rows of G.

Apparently the permuted code is a subcode of the original code. Since the two codes are equivalent they are identical. This completes the proof.

Thanks to (4.4.8) the Theorems (4.4.3) to (4.4.5) will now become very useful. The group PSL (2,n) is doubly transitive. Just as in (4.1.11) it follows from (4.1.10) that the minimum weight of the QR-code with generator $g_0(x)$ is odd. Therefore (4.4.3) and (4.4.4) yield:

(4.4.9) <u>THEOREM</u>: <u>The minimum weight of the QR code of length</u> n <u>with generator</u> $g_0(x)$ <u>is an odd number</u> d <u>for which</u>

(i) $d^2 > n$ <u>if</u> $n \equiv 1 \pmod 8$,

(ii) $d^2 - d + 1 \geq n$ <u>if</u> $n \equiv -1 \pmod 8$.

(4.4.10) <u>Example</u>: Over GF(2) we have

$$x^{23} - 1 = (x-1)(x^{11}+x^9+x^7+x^6+x^5+x+1)(x^{11}+x^{10}+x^6+x^5+x^4+x^2+1) =$$
$$= (x-1)g_0(x)g_1(x).$$

The binary QR code of length 23 with generator $g_0(x)$ is a (23,12) code. By (4.4.9) the minimum distance d of this code satisfies $d(d-1) \geq 22$. Since d is odd we have $d \geq 7$. The code is at least 3-error-correcting. Now the number of words in a sphere of radius 3 around a code word is $\sum_{i=0}^{3} \binom{23}{i} = 2^{11}$. It follows that the code has minimum distance 7 and that the code is perfect! This code is known as the <u>binary Golay code</u>. If we had applied (4.1.2) to find the minimum distance of this code we would have found $d \geq 5$.

(4.4.11) Example: Consider the binary QR-code of length 47. By (4.4.9) the minimum distance d of this code is at least 9. Now we apply (4.4.5) to find $d \geq 11$. Since $\sum_{i=0}^{6} \binom{47}{i} > 2^{23}$ the code cannot be 6-error-correcting, i.e. $d = 11$.

Several QR-codes are better than BCH codes but the decoding of these codes is not as easy. We shall not discuss decoding methods at all. The interested reader is referred to F. J. MacWilliams, Permutation Decoding of Systematic Codes, Bell System Tech. J. 43 (1964), 485-505.

4.5 Problems

(4.5.1) Let α be a primitive element of $GF(3^3)$ satisfying $\alpha^3 = \alpha - 1$. What is the generator of the 2-error-correcting ternary BCH code of length 26 if we use α as the primitive 26-th root of unity in (4.1.1)?

(4.5.2) What is the dimension of the ternary 5-error-correcting BCH code of length 80?

(4.5.3) Let α be a zero of $x^5 + x^2 + 1$ in $GF(2^5)$. We use α to define a binary 2-error-correcting BCH code of length 31. If a received word has the syndrome (1110011101) find the positions where errors have been made.

(4.5.4) If an (n,k) optimal code over GF(q) is a perfect 2-error-correcting code and $k > 1$, then $e = 1$. Prove this by considering the number of code words of weight $2e + 1$.

(4.5.5) Prove that the 2-nd order GRM code of length 25 over GF(5) has dimension 6 and minimum distance $d = 15$.

(4.5.6) Show that the ternary QR code of length 11 with generator $g_0(x)$ has minimum distance 5. Prove that this code is perfect. Compare with (3.5.6).

V. PERFECT CODES

5.1 Perfect single-error-correcting codes

The concept "perfect code" was introduced in (2.2.3) and then in (2.2.4) it was shown that Hamming codes over GF(q) are perfect. We recall that these codes were defined as follows: Let H be the m by $n := \frac{q^m - 1}{q - 1}$ matrix consisting of all the different nonzero columns with elements from GF(q) such that the first nonzero entry in each column is 1. Since no linear combination of two columns can be $\underline{0}$ we see that H is the parity check matrix of a linear (n,k) code over GF(q) with minimum distance 3. Because $q^k\{1 + n(q-1)\} = q^{m+k} = q^n$ the code is perfect.

In (3.2.1) it was shown that binary Hamming codes are equivalent to cyclic codes. This is not true for all Hamming codes but it is possible to generalize (3.2.1). Let $n := \frac{q^m - 1}{q - 1}$ and let α be a primitive element in $GF(q^m)$. If we define $\beta := \alpha^{q-1}$ then β is a primitive n-th root of unity. Now let V be the cyclic code of length n with as generator g(x) the minimal polynomial of β. The degree of g(x) is $\leq m$, i.e. the dimension of V is $\geq n - m$. If the code is single-error-correcting it is again perfect because $\{1 + n(q-1)\}q^{n-m} = q^n$. To show that the minimum distance of the code is at least 3 we must prove that for $0 < e < n$ and $a \in GF(q)$ the generator g(x) does not divide $x^e - a$. This is the case if $\beta^e \notin GF(q)$ for $0 < e < n$, i.e. $\beta^{e(q-1)} \neq 1$ for $0 < e < n$. Therefore it is necessary and sufficient to require that $(n,q-1) = 1$ which is true iff $(m,q-1) = 1$. The parity-check matrix of this cyclic code is $(1,\beta,\beta^2,\ldots,\beta^{n-1})$ which is a m by n matrix over GF(q) for which every pair of columns is linearly independent just as we had for the matrix H above.

From this discussion it follows that the first method of generalizing the binary Hamming codes yields more perfect codes than the method using cyclic codes. The smallest example where the cyclic-code approach fails is $q = 3$, $m = 2$ in which case $n = 4$. It is easily seen that the (4,2) ternary Hamming code is not equivalent to a cyclic code (see problem (1.5.7)).

Now we turn to the question whether there are other perfect single-error-correcting codes than the ones mentioned above. Let us consider an alphabet of q symbols where q is no longer necessarily a power of a prime but an arbitrary integer. Consider the set $\mathcal{R}^{(n)}$ of all n-letter words with symbols from this alphabet and let $V \subset \mathcal{R}^{(n)}$ be a perfect single-error-correcting code. This implies that the total number of words, i.e. q^n, is a multiple of the number of words in a sphere of radius 1 which is $1 + n(q-1)$. Therefore we have

(5.1.1) THEOREM. A necessary condition for the existence of a perfect single-error-correcting code of length n over an alphabet of q symbols is

$$[1 + n(q-1)] \mid q^n .$$

If $q = p^\alpha$ is a power of a prime this implies $1 + n(q-1) = q^a$.

Proof: We only have to show the last part. If $1 + n(q-1) = p^\lambda q^a$ with $0 \le \lambda < \alpha$ then we have

$$n = \frac{p^\lambda q^a - 1}{q-1} = p^\lambda \frac{q^a - 1}{q-1} + \frac{p^\lambda - 1}{q-1} .$$

The first term on the right-hand side is an integer. The second term is an integer only if $\lambda = 0$.

The only case where q is not a power of a prime for which it is known if a perfect code exists is $n = 7$, $q = 6$. The following proof is due to R. E. Block and M. Hall. Let V be a perfect single-error-correcting code of length 7 over the alphabet of 6 symbols (which we shall denote by 1,2,3,4,5,6). The code V has 6^5 words $(c_1,c_2,\dots,c_7)$ and no two of these can have the same initial 5-letter word $(c_1,c_2,\dots,c_5)$ since the minimum distance of V is 3. There are exactly 6^5 possible 5-tuples $(c_1,c_2,\dots,c_5)$ and therefore each of these is the initial 5-tuple of one code word of V. Therefore V contains 36 words of type $(1,1,1,i,j,a_{ij},b_{ij})$ and for these 36 words all pairs (i,j) are distinct. Now define the two 6×6 matrices A and B by $A := [a_{ij}]$, $B := [b_{ij}]$ $(i = 1,\dots,6;\ j = 1,\dots,6)$. Since the 36 words $(1,1,1,i,j,a_{ij},b_{ij})$ have distance ≥ 3 no row or column of A and B can contain the same symbol twice. Furthermore all pairs (a_{ij},b_{ij}) are distinct. Therefore A and B are two orthogonal Latin squares of order 6. It is known that two such Latin squares do not exist (cf. H. J. Ryser, Combinatorial Mathematics, p. 84). This proves:

(5.1.2) THEOREM: A perfect single-error-correcting code of length 7 over a 6-symbol alphabet does not exist.

Since a pair of orthogonal Latin squares of order q can be found for $q \neq 2$ or 6 we cannot generalize the proof of (5.1.2). The reasoning in the following proof resembles that of (5.1.2).

(5.1.3) THEOREM: If a perfect single-error-correcting code of length $n = q + 1$ over an alphabet of q symbols is a group code then q is a power of a prime.

Proof: Let $\mathcal{R}^{(n)}$ be the direct product of n copies of the abelian group G of order q. Just as in the proof of (5.1.2) we see that in a perfect code $V \subset \mathcal{R}^{(n)}$ each (q-1)-tuple $(c_1, c_2, \ldots, c_{q-1})$ occurs exactly once as the initial (q-1)-tuple of a code word. For $i = 1, 2, \ldots, q-1$ we define the mappings $\sigma_i: G \to G$ and $\tau_i: G \to G$ as follows. For $a \in G$ there is one code word $(c_1, c_2, \ldots, c_{q+1})$ with $c_i = a$ and $c_j = 0$ for $j \neq i$, $1 \leq j \leq q-1$. We define $\sigma_i(a) := c_q$, $\tau_i(a) := c_{q+1}$. Clearly σ_i and τ_i are permutations of G which fix 0. Since V is a subgroup of $\mathcal{R}^{(n)}$ we have $\forall_{a \in G} \forall_{b \in G} [\sigma_i(a+b) = \sigma_i(a) + \sigma_i(b)]$ and an analogous relation for τ_i. Therefore σ_i and τ_i are automorphisms of the group G. The code with words $(\sigma_1(c_1), \sigma_2(c_2), \ldots, \sigma_{q-1}(c_{q-1}), c_q, c_{q+1})$ is also a perfect group code V′. This code V′ has code words $\underline{c}(i,a) := (0,0,\ldots,0,a,0,\ldots,0,a,\tau_i(a))$ where the first a is in position i $(1 \leq i \leq q-1)$ and τ_i is an automorphism of G. For $a \in G$ and $1 \leq i_1 < i_2 \leq q-1$ the words $\underline{c}(i_1,a)$ and $\underline{c}(i_2,a)$ have distance 3 only if $\tau_{i_1}(a) \neq \tau_{i_2}(a)$. Now let p be a prime $p|q$ and let $a \in G$ be an element of order p in G. Since V′ is a group code the q-1 distinct elements $\tau_i(a)$ $(1 \leq i \leq q-1)$ also have order p, i.e. every nonzero element of G has order p. Therefore q is a power of the prime p.

The proof of (5.1.3) is due to B. Lindström. The theorem shows that if a perfect single-error-correcting code of length q+1 over an alphabet of q elements, where q is not a prime-power, exists then the code will not have anything like the algebraic structure of the codes we have studied in this course.

A method of showing non-existence of certain codes which sometimes leads to success is the following. If one can find the weight enumerator of the code (which supposedly exists) from the properties of the code and if at least one of the coefficients of this polynomial is not a non-negative integer then clearly the code cannot exist. An example of this method is (4.2.9). Although the method is going to fail in this section we shall nevertheless illustrate the attempt.

Let us assume that a perfect code V of length n over an alphabet of q symbols exists. Without loss of generality we can take these symbols to be 0, 1, ..., q-1 and assume that $(0,0,\ldots,0) \in V$. We define Hamming-weight in the usual way. In a sphere of radius 1 around a code word of weight j there are j words of weight j-1, $(q-2)j+1$ words of weight j and $(n-j)(q-1)$ words of weight j+1. Since V is perfect the union of all such spheres is $\mathcal{R}^{(n)}$. If $A(z) := \sum_{i=0}^{n} A_j z^j$ is the weight enumerator of V we must have

$$(5.1.4) \qquad (q-1)(n-j+1)A_{j-1} + [(q-2)j + 1]A_j + (j+1)A_{j+1} = \binom{n}{j}(q-1)^j$$

for $j = 0, 1, \ldots, n$. If we multiply (5.1.4) by z^j and sum over j we find

$$(5.1.5) \quad (q-1)nzA(z) - (q-1)z^2A'(z) + (q-2)zA'(z) + A(z) + A'(z) = [1+(q-1)z]^n.$$

The solution of (5.1.5) is

$$(5.1.6) \qquad A(z) = \frac{1}{n(q-1)+1}\left\{[1 + (q-1)z]^n + n(q-1)[1 + (q-1)z]^{\frac{n-1}{q}}(1-z)^{\frac{n(q-1)+1}{q}}\right\}.$$

We remark that this must then be the solution to (2.6.7) for the case $q = p^\alpha$. If q is not a power of a prime the substitution $n = \frac{q^m-1}{q-1}$ in (5.1.6) yields a polynomial with integral coefficients. This discourages us from trying to use (5.1.6) to show non-existence of perfect single-error-correcting codes.

We now turn to another question raised by (5.1.3) namely whether there are <u>non-linear perfect codes</u>. To make this problem meaningful we must identify equivalent codes and extend equivalence as defined in Chapter 2. If π is a permutation of the symbols of the alphabet we are considering and V is a code of length n then the code obtained by applying π to the i-th symbol $(1 \leq i \leq n)$ of the code words is called equivalent to V in the following discussion. We shall show that the answer to the question is positive. The method due to J. L. Vasiliev was generalized by J. Schönheim (On linear and nonlinear single-error-correcting q-nary perfect codes, Inf. and Control 12 (1968), 23-26) and B. Lindström (On group and nongroup perfect codes, Math. Scand. 25 (1969), 149-158.)

Let q be a power of a prime, $n = \frac{q^m - 1}{q - 1}$. Let $V \subset \mathcal{R}^{(n)}$ be a perfect, linear, single-error-correcting code. Define

$N := nq+1 = \frac{q^{m+1} - 1}{q - 1}$. Let $f\colon V \to GF(q)$ be any function with $f(\underline{0}) = 0$. Now order the nonzero elements of GF(q) in any way, i.e. $GF(q)\setminus\{0\} = \{\alpha_1, \alpha_2, \ldots, \alpha_{q-1}\}$. Define $p\colon (\mathcal{R}^{(n)})^{q-1} \to GF(q)$ by

$$p(\underline{v}_1, \underline{v}_2, \ldots, \underline{v}_{q-1}) := \sum_{i=1}^{q-1} \alpha_i \sum_{j=1}^{n} v_{ij} .$$

Now let $V^* \subset \mathcal{R}^{(N)}$ be defined by

$$V^* := \Big\{(\underline{v}_1, \underline{v}_2, \ldots, \underline{v}_{q-1}, \underline{c} + \sum_{i=1}^{q-1} \underline{v}_i, p(\underline{v}_1, \underline{v}_2, \ldots, \underline{v}_{q-1}) + f(\underline{c})) \;\Big|\; \underline{v}_i \in \mathcal{R}^{(n)} \text{ for } i = 1, 2, \ldots, q-1;\ \underline{c} \in V\Big\}$$

(where the notation is obvious). This is a subset of $(q^n)^{(q-1)} q^{n-m} = q^{N-(m+1)}$ elements of $\mathcal{R}^{(N)}$. To show that V^* is also a perfect single-error-correcting code it is sufficient to show that any two vectors in V^* have distance ≥ 3. Let

$$\underline{u} := (\underline{v}_1, \underline{v}_2, \ldots, \underline{v}_{q-1}, \underline{c} + \sum_{i=1}^{q-1} \underline{v}_i, p(\underline{v}_1, \ldots, \underline{v}_{q-1}) + f(\underline{c}))$$

$$\underline{u}' := (\underline{v}_1', \underline{v}_2', \ldots, \underline{v}_{q-1}', \underline{c}' + \sum_{i=1}^{q-1} v_i', p(\underline{v}_1', \ldots, \underline{v}_{q-1}') + f(\underline{c}'))$$

be two different vectors in V^*. We write the distance $d(\underline{u}, \underline{u}')$ as the sum of three integers d_1, d_2, d_3 namely the contribution of the first $n(q-1)$ components to $d(\underline{u}, \underline{u}')$, the contribution of the next n components and finally the contribution of the last component. We consider several cases:

(a) $d_1 \geq 3$. Nothing to prove.

(b) If $d_1 = 0$ then $\underline{c} \neq \underline{c}'$ and hence $d_2 \geq 3$ because $\underline{c}$ and $\underline{c}'$ are both in V.

(c) If $d_1 = 1$ and $\underline{c} \neq \underline{c}'$ then d_2 is at least 2 because $\underline{c} - \underline{c}' \in V$. Therefore $d(\underline{u},\underline{u}') = d_1 + d_2 + c_3 \geq 3$.

(d) If $d_1 = 1$ and $\underline{c} = \underline{c}'$ then $d_2 = 1$ and $p(\underline{v}_1,\dots,\underline{v}_{q-1}) \neq p(\underline{v}_1',\dots,\underline{v}_{q-1}')$, i.e. $d_3 = 1$. Hence $d_1 + d_2 + d_3 = 3$.

(e) For some k, $1 \leq k \leq q-1$, the vectors $\underline{v}_k$ and $\underline{v}_k'$ have distance 2 and for all other i in $1 \leq i \leq q-1$ the vectors $\underline{v}_i$ and $\underline{v}_i'$ are identical. Then $d_1 = 2$. Again using $\underline{c} - \underline{c}' \in V$ we see that $d_2 \geq 1$ and hence $d(\underline{u},\underline{u}) \geq 3$.

(f) For a pair i,j with $1 \leq i < j \leq q-1$ we have $d(\underline{v}_i,\underline{v}_i') = 1$, $d(\underline{v}_j,\underline{v}_j') = 1$ and for all other k in $1 \leq k \leq q-1$ we have $\underline{v}_k = \underline{v}_k'$. Then again $d_1 = 2$ and again $d_2 \geq 1$ if $\underline{c} \neq \underline{c}'$. The only difficulty arises if $\underline{c} = \underline{c}'$. We must now check if $d_2 = d_3 = 0$ is possible. Let $v_{i\mu} \neq v_{i\mu}'$, $v_{j\nu} \neq v_{j\nu}'$. If $d_2 = 0$ then $\mu = \nu$ and $v_{i\mu} + v_{j\mu} = v_{i\mu}' + v_{j\mu}'$. But then

$$\begin{aligned} p(\underline{v}_1,\dots,\underline{v}_{q-1}) - p(\underline{v}_1',\dots,\underline{v}_{q-1}') &= \alpha_i(v_{i\mu} - v_{i\mu}') + \alpha_j(v_{j\mu} - v_{j\mu}') \\ &= (\alpha_i - \alpha_j)(v_{i\mu} - v_{i\mu}') \neq 0, \end{aligned}$$

i.e. $d_3 = 1$.

The cases (a) to (f) cover all possibilities. Clearly $\underline{0} \in V^*$. If we choose f in such a way that it does not take all the values in its range the same number of times then V^* cannot be a linear subspace of $\mathcal{R}^{(N)}$. To be able to choose f in such a way the code V must have at least 3 code words which is the case if $q = 2$ and $m > 2$ or $q \geq 3$ and $m \neq 1$. Therefore we have proved

(5.1.7) <u>THEOREM</u>: <u>If</u> q <u>is a power of a prime</u>, $N := \frac{q^{m+1} - 1}{q - 1}$, <u>where</u> $m \geq 3$ <u>if</u> $q = 2$ <u>and</u> $m \geq 2$ <u>if</u> $q \geq 3$, <u>then there exists a perfect single-error-correcting code of length</u> N <u>over</u> GF(q) <u>which is not equivalent to a linear code</u>.

In Section (2.2) we mentioned the connection between ternary Hamming codes and the football-pool problem. In a more general form the problem was proposed by O. Taussky and J. Todd (Ann. Soc. Polon. Math. 21 (1948), 303-305):

(5.1.8) PROBLEM: Let G be an abelian group with n base elements $g_1, g_2, \ldots, g_n$ of order p (corresponding to $\mathcal{R}^{(n)}$ in our notation). Let S be the set $\{g_i^k \mid i = 1,\ldots,n;\ k = 0,1,\ldots,p-1\}$ (corresponding to our concept of a sphere with radius 1). Determine the minimal integer $\sigma(n,p)$ such that there exists a subset $H \subset G$ with $\sigma(n,p)$ elements such that $G = HS$. When is it possible to take for H a subgroup of G?

It was shown by B. Kuttner, J. G. Mauldon and S. K Zaremba that if p is a power of a prime and $|S|$ a power of the same prime then the problem has a solution where H is a subgroup and the decomposition of G unique. Clearly this amounts to saying that a perfect single-error-correcting code exists. In the formulation of (5.1.8) the problem is called the covering-problem for groups. In the case of a perfect covering we have a perfect code. The general covering problem is a hard combinatorial problem and only few of the values $\sigma(n,p)$ are known.

5.2 The sphere-packing condition

Let us now turn to the harder problem of finding perfect e-error-correcting codes, with $e \geq 2$, over an alphabet of q symbols. We shall denote the block length by n as usual. If q is a power of a prime we write $q = p^\alpha$.

Now the number of words in a sphere of radius e around any code word is $1 + \binom{n}{1}(q-1) + \binom{n}{2}(q-1)^2 + \cdots + \binom{n}{e}(q-1)^e$. As an analog of (5.1.1) we find the following sphere-packing condition:

(5.2.1) THEOREM: A necessary condition for the existence of a perfect e-error-correcting code of block length n over an alphabet of q symbols is

$$\sum_{i=0}^{e} \binom{n}{i}(q-1)^i \;\Big|\; q^n .$$

(5.2.2) COROLLARY: If $q = p^{\alpha}$ the condition in (5.1.1) can be stated as

$$\sum_{i=0}^{e} \binom{n}{i}(q-1)^i = q^k$$

for some integer k.

Proof: By (5.1.1) we have $\sum_{i=0}^{e} \binom{n}{i}(q-1)^i = p^{\beta}q^k$ with $0 \leq \beta < \alpha$ and some integer $k \geq 0$. Also $\sum_{i=0}^{n} \binom{n}{i}(q-1)^i = q^n$. By subtracting we find

$$q^n - p^{\beta}q^k = \sum_{i=e+1}^{n} \binom{n}{i}(q-1)^i \equiv 0 \pmod{q-1},$$

i.e.

$$q^k \left\{(q^{n-k}-1) - (p^{\beta}-1)\right\} \equiv 0 \pmod{q-1}$$

which implies $\beta = 0$. (Compare with the proof of (5.1.1)).

Remark: We could have used the same proof as in (5.1.1). This proof has the advantage that it shows that $q^{n-k} - 1 \equiv 0 \bmod (q-1)^{e+1}$. Also remark that if $n > e$ the left-hand side in (5.2.2) is more than $\sum_{i=0}^{e} \binom{e}{i}(q-1)^i = q^e$, i.e. $k > e$.

(5.2.3) COROLLARY: If $q = p^{\alpha}$ the condition in (5.1.1) can be stated as

$$\sum_{j=0}^{e} (-1)^j q^j \binom{n}{j}\binom{n-j-1}{e-j} = (-1)^e q^k.$$

Proof: We start from (5.2.2) and in the sum on the left-hand side expand $(q-1)^i$ in powers of q and then rearrange. The result follows from

$$\sum_{i=j}^{e} (-1)^i \binom{n}{i}\binom{i}{j} = \sum_{i=j}^{e} (-1)^i \binom{n}{j}\binom{n-j}{i-j} = \binom{n}{j} \sum_{k=0}^{e-j} (-1)^{k+j}\binom{n-j}{k} =$$

$$= (-1)^e \binom{n}{j}\binom{n-j-1}{e-j}.$$

Before taking a closer look at the equation (5.1.2) we shall obtain one more result which is a straightforward consequence of the perfect sphere-packing. Without loss of generality we may assume that (0,0,...,0) is in the perfect code. Now every word of weight e + 1 is not in the sphere $S_e(\underline{0})$ but must be in exactly one sphere of radius e around a code word. Since the minimum weight of the code is 2e+1 each word of weight e+1 must be in a sphere $S_e(\underline{x})$ where $w(\underline{x}) = 2e + 1$. On the other hand, such a sphere contains $\binom{2e+1}{e}$ words of weight e + 1. It follows that the code contains

$$a_{2e+1} := \frac{\binom{n}{e+1}(q-1)^{e+1}}{\binom{2e+1}{e}}$$

words of weight e (cf. problem 4.5.4.). Therefore we have:

(5.2.4) <u>THEOREM</u>: <u>A necessary condition for the existence of a perfect</u> e <u>-error-correcting code of block length</u> n <u>over an alphabet of</u> q <u>symbols is that</u>

$$(q-1)^{e+1}\binom{n}{e+1}/\binom{2e+1}{e}$$

<u>is an integer</u>.

The necessary condition of (5.2.4) is one of a sequence which we shall now derive using a generalization of the concept of block design.

(5.2.5) DEFINITION: A <u>tactical configuration</u> of type λ; t-d-n (also called a t-<u>design</u>) is a collection $\mathcal{D}$ of d-subsets of an n-set S such that every t-subset of S is contained in exactly λ distinct members of $\mathcal{D}$. To avoid trivial configurations one usually demands $0 < t < d < n$.

For the special case t = 2 the tactical configuration is called a balanced incomplete block design (cf. H. J. Ryser, Combinatorial Mathematics). In the case λ = 1 the tactical configuration is called a <u>Steiner system</u> of type (t,d,n). The study of different kinds of tactical configurations, a.o. finite projective planes, is an important part of combinatorial analysis. The connection with perfect codes is established by the following theorem. We restrict the theorem to binary codes. The generalization to $q = p^\alpha$ is possible by adding the assumption of linearity.

(5.2.6) THEOREM: Let V be a binary perfect e-error-correcting code of block length n and assume $\underline{0} \in V$. We consider the collection $\mathcal{D}$ of $(2e+1)$-subsets D of $\{1,2,\dots,n\}$ for which there is a code word $\underline{c} \in V$ of weight $2e+1$ with its nonzero coordinates in the positions of D. This is a Steiner system of type $(e+1,\ 2e+1,\ n)$.

Proof: The proof is exactly the same as the proof of (5.2.4).

Now if a Steiner-system $\mathcal{D}$ of type $(t,\ d,\ n)$ exists with $S = \{1,2,\dots,n\}$ take any h-subset of S, e.g. $H := \{1,2,\dots,h\}$, where $0 \le h \le t$. We now count the t-subsets of S containing H. This number is obviously $\binom{n-h}{t-h}$. Each of these t-subsets is contained in exactly one d-subset belonging to $\mathcal{D}$. Every d-subset in $\mathcal{D}$ which contains H contains exactly $\binom{d-h}{t-h}$ t-subsets which also contain H. Therefore the number $\binom{n-h}{t-h}$ must be a multiple of $\binom{d-h}{t-h}$. Combined with (5.2.6) this proves:

(5.2.7) THEOREM: If a binary perfect e-error-correcting code of block length n exists then the numbers

$$\binom{n-h}{e+1-h} \Big/ \binom{2e+1-h}{e+1-h}, \qquad h = 0,1,\dots,e$$

are all integers.

Note that substitution of $h = 0$ in (5.2.7) gives us (5.2.4) again. Substitution of $h = e$ yields

(5.2.8) COROLLARY: If a binary perfect e-error-correcting code exists then $\frac{n+1}{e+1}$ is an integer.

Extensive computer searches have been made to find solutions of the equation of (5.2.2) with $e > 1$. The ranges that were covered were:

(1) $e = 2$; q odd; $3 \le q \le 125$, $3 \le k \le 40000$ (E. L. Cohen 1964),

(2) $e \le 20$; $q = 2$; $n \le 2^{70}$ (M. H. McAndrew 1965),

(3) $e \le 1000$; $q \le 100$; $n \le 1000$ (J. H. vanLint 1967).

No other solutions were found than trivial ones (e.g. $e = n$) and,

(a) $q = 2$, $n = 2e + 1$ corresponding to repetition codes,

(b) $q = 2$, $e = 2$, $n = 90$ (excluded by (5.2.8)),

(c) $q = 2$, $e = 3$, $n = 23$ corresponding to the binary Golay code treated in (4.4.10),

(d) $q = 3$, $e = 2$, $n = 1$ corresponding to the ternary Golay code treated in problem (3.5.6).

(See E. L. Cohen, A note on perfect double error correcting codes on q symbols, Inf. and Control 7 (1964), 381-384; M. H. McAndrew, An algorithm for solving a polynomic congruence and its application to error-correcting codes, Math. of Comp. 19 (1965), 68-72; [9]).

Of course computer searches cover only a finite range. We shall show two examples of an infinite number of cases ruled out by the condition (5.2.1).

We consider the condition (5.2.2) for $q = 2$ and e odd. We shall prove

(5.2.9) LEMMA: If e is odd then

$$\sum_{i=0}^{e} \binom{n}{i} = \frac{1}{e!} (n+1) R_e(n)$$

where $R_e(n)$ is a polynomial in n of degree $e - 1$ with integral coefficients.

Proof: Clearly $\sum_{i=0}^{1} \binom{n}{i} = n + 1 = \frac{1}{1!}(n+1)R_1(n)$ with $R_1(n) = 1$. We proceed by induction:

$$\binom{n}{e+1} + \binom{n}{e+2} = \frac{1}{(e+2)!} \{(e+2) + (n-e-1)\} \prod_{i=0}^{e} (n-i),$$

i.e.

$$R_{e+2}(n) = (e+2)(e+1)R_e(n) + \prod_{i=0}^{e} (n-i)$$

which proves the lemma.

We now show one example of how this lemma can be used (cf. H. S. Shapiro and D. S. Slotnick, On the mathematical theory of error-correcting codes, IBM. J. Res.

Develop. 3 (1959), 25-34).

If a binary perfect 3-error-correcting code exists then by (5.2.2) and (5.2.9) we have

$$(n+1)(n^2-n+6) = 2^{k+1} \cdot 3$$

or

$$(n+1)\{(n+1)^2 - 3(n+1) + 8\} = 2^{k+1} \cdot 3.$$

If $(n+1)$ is divisible by 16 then the highest power of 2 which divides $n^2 - n + 6$ is 2^3, i.e. $n^2 - n + 6$ is a divisor of 24 and then $n + 1 < 16$. Therefore $(n+1)$ is not divisible by 16 and hence $(n+1)$ is a divisor of 24. This leaves only the following possible values for n:

(a) $n = 0,1,2$ not corresponding to codes,

(b) $n = 3$ corresponding to the trivial one-word code of length 3,

(c) $n = 7$ corresponding to the (trivial) repetition code,

(d) $n = 23$ corresponding to the binary Golay code.

This proves:

(5.2.10) <u>THEOREM</u>: <u>The binary Golay code is the only non-trivial binary perfect 3-error-correcting code.</u>

By Lemma (5.2.9) we can do the same type of calculation for other odd e. This has been done for $e < 20$ yielding no new binary perfect codes. Of course the first case one would naturally study is $e = 2$, $q = 2$. This is more difficult than (5.2.10). In this case we have from (5.2.2):

$$(2n+1)^2 = 2^{k+3} - 7.$$

The equation $x^2 + 7 = 2^m$ has been studied in several papers (cf. Math. Rev. 26, #74). The methods depend on unique factorization in $\mathbb{Q}(\sqrt{-7})$. The only solutions are $x = 1, 3, 5, 11$ and 181 corresponding to $n = 0, 1$ (no codes), $n = 2$ (trivial code), $n = 5$ (repetition code), $n = 90$ (excluded by (5.2.8)). Therefore:

(5.2.11) <u>THEOREM</u>: <u>There is no non-trivial binary perfect 2-error-correctiong code.</u>

In Section (5.5) we shall prove this in a simple way. We give one more example. Let $q = 6$ and $e = 2$. If a perfect code of block length n with these parameters exists, then by (5.2.1)

$$1 + 5\binom{n}{1} + 25\binom{n}{2} = 2^a 3^b,$$

i.e.

$$(5n-1)(5n-2) = 2^{a+1} 3^b.$$

Since the factors on the left-hand side are relatively prime one must be a power of 2 and the other a power of 3. It is known that $|2^{a+1} - 3^b| = 1$ has as only solutions the pairs (2,1), (2,3), (4,3), and (8,9) which give $n = 1$ and $n = 2$ as only solutions of which only $n = 2$ corresponds to a code namely the trivial one-word code.

5.3 The Golay codes

In (4.4.10) we introduced the binary (23,12) <u>Golay code</u> which is a perfect 3-error-correcting linear code. By (5.2.6) the code words of weight 7 in the Golay code form a Steiner system of type (4,7,23). We now consider the extended code which is a (24,12) code. By (4.4.8) this code is invariant under the doubly transitive group PSL (2,23). The group of automorphisms of the code is in fact much larger. It is the <u>Mathieu group</u> M_{24} which is 5-fold transitive. Because of the interesting properties of this group the Golay codes (and possible other perfect codes!) have become objects of interest to group theorists. We shall not discuss M_{24} here (cf. e.g. E. F. Assmus and H. F. Mattson, Perfect codes and the Mathieu groups, Archiv der Math. <u>17</u> (1966), pp. 121-135).

Let S be a 5-subset of the positions of the code words i.e. $\{0,1,\ldots,22;\infty\}$. Clearly at most one code word of the extended Golay code has ones in the positions of S. If $\infty \in S$ then by (5.2.6) exactly one code word of weight 8 in the extended Golay code has ones in the positions of S. If $\infty \notin S$ then there must be exactly one code word of the Golay code with weight ≤ 8 and ones in the positions of S and hence exactly one word of weight 8 in the extended Golay code which has ones in the positions of S. We have proved:

(5.3.1) <u>THEOREM</u>: <u>The code words of weight 8 in the extended binary Golay code form a Steiner system of type (5,8,24).</u>

(5.3.2) COROLLARY: Let $\lambda_2 := 77$, $\lambda_3 := 21$, $\lambda_4 := 5$. Then for $i = 2,3,4$ and for every i-subset S of $\{0,1,\dots,22;\infty\}$ there are λ_i code words of weight 8 in the extended binary Golay code such that these code words have ones in the positions of S.

<u>Proof</u>: Just as in (5.2.7) we have $\lambda_i = \binom{24-i}{5-i} / \binom{8-i}{5-i}$.

In connection with (5.3.1) we remark that there are no known t-designs with $t > 5$.

For the binary extended Golay code the following generalization of the idea of threshold decoding as treated in Section 2.4 was suggested by J. M Goethals (cf. On the Golay perfect binary code, Report R93, MBLE, Brussels). The decoding method depends on (5.3.2) and the following theorem:

(5.3.3) <u>THEOREM</u>: <u>The extended binary Golay code is self dual.</u>

<u>Proof</u>: Let $(c_0, c_1, \dots, c_{22}, c_\infty)$ and $(d_0, d_1, \dots, d_{22}, d_\infty)$ be two code words. Just as in the proof of (4.4.4) the product $(\sum_{j=0}^{22} c_i x^i)(\sum_{j=0}^{22} d_j x^{-j})$ is 0 or $1 + x + \dots + x^{22}$, the latter iff $\Sigma c_i = \Sigma d_i = 1$ which is the case if $c_\infty = d_\infty = 1$. It follows that $\sum_{i=0}^{22} c_i d_i + c_\infty d_\infty = 0$.

The decoding method now works as follows. The extended binary Golay code has 253 code words of weight 8 with a 1 in a specified position (cf. (5.2.4)). By (5.3.3) these 253 code words can be used as parity-checks. If the result of substitution of a received word in a parity-check equation is 1 we shall call this a parity-check failure. From (5.3.2) we then find the following table:

Number of parity-check failures for the 253 p.c. equations

# of errors outside the fixed position →	0	1	2	3	4
fixed position correct	0	77	112	125	128
" " in error	253	176	141	128	125

From the table we see that if the number of errors is ≤ 3 we can decide if the fixed position is correct or not. Then using the fact that the Golay code is cyclic the errors can be corrected successively. In this procedure the number of errors is known after the first step. The numbers in the table show that 4 errors are detected but not corrected (we already knew this). Another threshold-type decoding procedure was suggested by E. F. Assmus and H. F. Mattson (Report AFCRL-68-0478 of the Applied Research Laboratory of Sylvania Electronic Systems). It also depends on (5.3.2) and (5.3.3).

It was shown by V. Pless (On the uniqueness of the Golay codes, J. Comb. Theory **5** (1968), 215-228) that the Golay codes are unique. Therefore every construction leading to a perfect binary (23,12) code gives us a code equivalent to the binary Golay code as introduced in (4.4.10). The following very simple construction is due to E. F. Assmus and H. F. Mattson.

In the description below we use $\underline{1}$ to designate the all-one vector and $|\underline{x}|$ for the Hamming-weight of $\underline{x}$. The product $\underline{u}\,\underline{v}$ of two vectors is defined as in (2.3.1). Let H be the extended (8,4) Hamming code. Using 3.1 and 3.3 (or simply by inspection) we see that H consists of $\underline{0}$, the seven cyclic shifts of (1,1,0,1,0,0,0) each followed by a parity-check 1 and the complements of these vectors, i.e. $\underline{1}$ and the seven shifts of (0,0,1,0,1,1,1) each with a 0-parity-check. Numbering the positions of H from 1 to 8 and applying the permutation (1,7)(2,6)(3,5) we obtain the equivalent code H′. We remark that $H \cap H' = \{\underline{0},\underline{1}\}$, that all other vectors of H and H′ have weight 4 and that all vectors of H + H′ have even weight. Now:

$$V := \{(\underline{a} + \underline{x},\ \underline{b} + \underline{x},\ \underline{a} + \underline{b} + \underline{x}) \mid \underline{a} \in H,\ \underline{b} \in H,\ \underline{x} \in H'\}$$

is clearly a (24,12) linear code. To show that V is the extended Golay code we only have to show that V has minimum distance 8. If $\underline{v} = (\underline{a}+\underline{x}, \underline{b}+\underline{x}, \underline{a}+\underline{b}+\underline{x}) \neq \underline{0}$ and at least one of the vectors $\underline{a}$, $\underline{b}$, $\underline{a}+\underline{b}$, $\underline{x}$ is either $\underline{0}$ or $\underline{1}$ then $|\underline{v}| \geq 8$. To check the other vectors $\underline{v}$ we use the equality:

$$|\underline{u} + \underline{v}| + 2|\underline{u}\,\underline{v}| = |\underline{u}| + |\underline{v}|.$$

Applying this three times we find

$$|\underline{a} + \underline{x}| + |\underline{b} + \underline{x}| + |\underline{a} + \underline{b} + \underline{x}| = |\underline{a} + \underline{b}| + 2|(\underline{a}+\underline{x})(\underline{b}+\underline{x})| + |\underline{a} + \underline{b} + \underline{x}| =$$

$$= |x| + 2\{|(\underline{a}+\underline{x})(\underline{b}+\underline{x})| + |(\underline{a}+\underline{b})(\underline{1}+\underline{x})|\} +$$

$$= |x| + 2|\underline{a} + \underline{b} + \underline{a}\,\underline{b} + \underline{x}|.$$

Since $|\underline{x}| = 4$ we must show that $|\underline{a} + \underline{b} + \underline{a}\,\underline{b} + \underline{x}| \geq 2$. Since H is self-dual $\underline{a}\,\underline{b}$ has even weight and it is therefore sufficient to show that $\underline{a} + \underline{b} + \underline{a}\,\underline{b} \neq x$. Now $\underline{a} + \underline{b} + \underline{a}\,\underline{b} = \underline{x}$ implies $(\underline{a}+\underline{1})(\underline{b}+\underline{1}) = \underline{x} + \underline{1}$ where $\underline{a} + \underline{1}$, $\underline{b} + \underline{1}$ and $\underline{x} + \underline{1}$ all have weight 4 which is only possible if $\underline{a} = \underline{b}$. Since $\underline{a} \neq \underline{x}$ this is impossible. If we finally shorten V we obtain the Golay code.

Yet another representation was given by M. Karlin. The quadratic residues mod 11 form an (11,5,2) difference set namely $\mathcal{D} := \{1,3,4,5,9\}$. (For definitions and proof cf. H. J. Ryser, Combinatorial Mathematics). Now let C be the circulant with first row (11011100010), i.e. ones in position 0,1,3,4,5,9. Then $CC^T = C^TC = 3I + 3J$ where I is the unit matrix and J the all-one matrix. Consider the cyclic code generated by the rows of C. Since $(1 + x + x^3 + x^4 + x^5 + x^9, x^{11} - 1) = x - 1$ this code is the 10-dimensional code consisting of all even-weight words, i.e. C has rank 10. Using all these properties it is not hard to check that the matrix

$$G = \left(I_{12} \;\middle|\; \begin{array}{c} 1111 \cdots 1 \\ \hline C \end{array} \right)$$

generates a (23,12) code with minimum weight 7 (remark that each row of G has weight 7 or 12). We omit the rather tedious details. Apparently G is the generator matrix in reduced echelon form of the binary Golay code.

The fact that the Golay code is perfect is enough to find the weight enumerator of this code. By combining several theorems which we have proved the weight enumerator can be found much faster. Let $\sum_{i=0}^{23} a_i z^i$ be the weight enumerator. We know from (5.2.4) that $a_7 = 253$. From (4.4.8) and (4.1.10) we find $a_8 = 506$. The total number of words of weight 7 which are at a distance ≤ 3 from the code words of weight 7 and 8 is:

$$a_7\{1 + \binom{7}{1}\binom{16}{1}\} + a_8\{\binom{8}{1} + \binom{8}{2}\binom{15}{1}\} = \binom{23}{7}.$$

Therefore the code has $a_9 = a_{10} = 0$. Since the Golay code contains the all-one vector we have $a_{23} = 1$, $a_{16} = 253$, $a_{15} = 506$, $a_{14} = a_{13} = 0$. From (4.1.10) we have $a_{11} = a_{12}$. Since $\Sigma a_i = 2^{12}$ we find $a_{11} = a_{12} = 1288$. The weight enumerator of the extended binary Golay code is

$$(1 + z^{24}) + 759(z^8 + z^{16}) + 2576z^{12}.$$

The ternary Golay code was first introduced in problem (3.5.6). Since then we have seen that this perfect 2-error-correcting code is a QR code. The methods of Section 4.4 allow us to prove that the code is perfect somewhat faster than was done in (3.5.6). Let $g_0(x) := x^5 + x^4 - x^3 + x^2 - 1$. If C^+ denotes the ternary Golay code (generator $g_0(x)$) and C the subcode with generator $(x-1)g_0(x)$ then we know from (4.4.4) that every code word in $C^+ \setminus C$ has weight ≥ 4. In the same way as (5.3.3) was proved we find that any two words in C have inner product 0. Now the generator $(x-1)g_0(x) = x^6 + x^4 - x^3 - x^2 - x + 1$ has weight 6 and the same of course holds for the cyclic shifts of the generator. Let $\underline{x}$ and $\underline{y}$ be two vectors of C with $|\underline{x}| \equiv |\underline{y}| \equiv 0 \pmod 3$. Let α denote the number of positions where $\underline{x}$ and $\underline{y}$ have the same nonzero coordinates and let β be the number of positions where $\underline{x}$ and $\underline{y}$ both have nonzero entries but different ones. Then $|\underline{x}| = \alpha + \beta + \gamma_1$, $|\underline{y}| = \alpha + \beta + \gamma_2$ and $0 = (\underline{x},\underline{y}) = \alpha - \beta$ and finally $|\underline{x} + \underline{y}| = \alpha + \gamma_1 + \gamma_2 \equiv |\underline{x}| + |\underline{y}| \equiv 0 \pmod 3$. From this it follows that all code words in C have a weight $\equiv 0 \pmod 3$. Furthermore no vector in C has weight 3 because if there were such a code word a cyclic shift would have the form $(1+x^a+x^b)$ with $0 < a < b \leq 10$ and then $(1+x^a+x^b)(1+x^{-a}+x^{-b}) = 0$ would imply $a \equiv -b \equiv b - a \pmod{11}$ which is impossible. It follows that C has minimum weight 6. Now in order to show that C^+ is perfect we only have to show that a vector in $C^+ \setminus C$ cannot have weight 4. If there were such a code word there would also be a code word of the form $v(x) = 1 + x^a + x^b \pm x^c$. Then $(1+x+x^2+\cdots+x^{10}) \pm v(x)$ would be a word of weight 11 or 8 in C which we have shown to be impossible.

The ternary Golay code is small enough to make decoding procedures of a complicated nature superfluous. A simple table of syndromes is just as fast.

For the ternary Golay code we can easily find a generating matrix in reduced echelon form of the same type as we had for the binary Golay code. Let C be the circulant with first row (0,1,-1,-1,1). Then define

$$G := \begin{pmatrix} I_6 & \begin{array}{c} 1\;1\;1\;1\;1 \\ \boxed{\quad C \quad} \end{array} \end{pmatrix},$$

$$H := \begin{pmatrix} \begin{array}{c} 1 \\ 1 \\ 1 \\ 1 \\ 1 \end{array} & C & -I_5 \end{pmatrix}.$$

Then G has rank 6, H has rank 5 and $GH^T = 0$. From

$$GG^T = \begin{pmatrix} 0 & 0\;0\;0\;0\;0 \\ \begin{array}{c} 0 \\ 0 \\ 0 \\ 0 \end{array} & -J_5 \end{pmatrix}$$

it follows that, with $\underline{a} := (a_1, a_2, \ldots, a_6)$:

$$\underline{a}\, GG^T \underline{a}^T = -\Big(\sum_{i=2}^{6} a_i\Big)^2 \neq 1,$$

i.e. all code words $\underline{a}$ G in the code with generator G (and parity check matrix H) have a weight $\not\equiv 1 \pmod 3$. The rows of G have weight 5 or 6. A linear combination of 3 or more rows of G has weight > 3, and hence ≥ 5. So all we have to do is to check the linear combinations of two rows of G by hand to see that G generates a ternary (11,6) code with minimum distance 5. This code must then be equivalent to the ternary Golay code.

The weight enumerator for this code was found in (3.5.6).

5.4 Lloyd's Theorem

In 1957 S. P. Lloyd ([12]) proved a theorem which gave a necessary condition for the existence of a binary perfect e-error-correcting code. This theorem was later generalized by F. J. MacWilliams and recast by A. M. Gleason into an algebraic form. The theorem did not receive the attention it deserves because it was hard to apply. In the next section we shall see how Lloyd's theorem was used recently to show the non-existence of a large class of perfect codes. We shall present Gleason's proof of Lloyd's theorem in this section.

(5.4.1) NOTATION: (i) $\mathcal{R} := \mathcal{R}^{(n)}$ will denote, as usual, the n-dimensional vector space over GF(q).

(ii) $\mathfrak{A}$ will be a vector space of dimension q^n over $\mathbb{Q}$ with the elements of $\mathcal{R}$ as basis vectors. We shall denote addition in $\mathfrak{A}$ by $\oplus$. The elements of $\mathfrak{A}$ are linear combinations $a_1\underline{v}_1 \oplus a_2\underline{v}_2 \oplus \cdots$ $(a_i \in \mathbb{Q},\ \underline{v}_i \in \mathcal{R})$. We shall denote such sums by $\bigoplus\limits_{\underline{v}_i \in \mathcal{R}} a_i\underline{v}_i$ to distinquish from the addition in the group $(\mathcal{R},+)$. We use capital letters for elements of $\mathfrak{A}$.

(iii) In $\mathfrak{A}$ we define a binary operation, denoted by the symbol $*$, by

$$\Big(\bigoplus_{\underline{v}_i \in \mathcal{R}} a_i\underline{v}_i\Big) * \Big(\bigoplus_{\underline{v}_j \in \mathcal{R}} b_j\underline{v}_j\Big) := \bigoplus_{\underline{v} \in \mathcal{R}} \Big(\sum_{\underline{v}_i+\underline{v}_j=\underline{v}} a_i b_j\Big)\, \underline{v}$$

which is just the <u>formal</u> product of the left-hand sides if we interpret $\underline{v}_i * \underline{v}_j$ as $\underline{v}_i + \underline{v}_j$.

We then have:

(5.4.2) <u>THEOREM</u>: $(\mathfrak{A}, \oplus, *)$ <u>is a ring</u>.

Proof: We leave this to the reader as an exercise.

The ring $\mathfrak{A}$ (which is also vector space over $\mathbb{Q}$) is called the group ring of $\mathcal{R}$ over $\mathbb{Q}$.

To simplify the notation we use the following convention: If $V \subset \mathcal{R}$ then we also use V to denote the element $\sum_{\underline{v} \in V} \underline{v}$ in $\mathfrak{A}$. As before, we use $w(\underline{v})$ to denote the weight of $\underline{v} \in \mathcal{R}$ and we use S_e to denote the sphere of radius e with center $\underline{0}$, i.e.

$$S_e := \{\underline{v} \in \mathcal{R} \mid w(\underline{v}) \leq e\}$$

and by the previous convention we also have

$$S_e = \sum_{\underline{v} \in \mathcal{R}, w(\underline{v}) \leq e} \underline{v} .$$

In the rest of the section we shall assume that a perfect e-error-correcting code $C \subset \mathcal{R}^{(n)}$ exists, where $e < n$, and without loss of generality we assume that $\underline{0} \in C$. By the definition of a perfect code and (5.4.1) (iii) we have

(5.4.3) $$C * S_e = \mathcal{R}.$$

(5.4.4) NOTATION: The element $(1,1,\ldots,1,0,0,\ldots,0) \in \mathcal{R}$ for which the first k components are 1 and the remaining components are 0 will be denoted by $\underline{v}_k$.

Obviously we have

(5.4.5) $$\mathcal{R} = \{\underline{v}_i\} * (S_e * C) = S_e * (\underline{v}_i * C) \quad \text{for } i = 0,1,\ldots,e.$$

Here the element $\underline{v}_i * C \in \mathfrak{A}$ corresponds to the set

$$C_i := \{\underline{c} + \underline{v}_i \mid \underline{c} \in C\} \subset \mathcal{R}$$

which is a perfect code obtained by translating C. Since C_i contains a word of weight i but no word of smaller weight $(i = 0,1,\ldots,e)$ we see that the elements $C_i \in \mathfrak{A}$ are linearly independent over $\mathbb{Q}$.

(5.4.6) DEFINITION: For $i = 0, 1, \ldots, n$ we define:

(i) $Y_i := \sum_{\underline{v} \in \mathfrak{R}, w(\underline{v})=i} \underline{v}.$

(ii) The <u>symmetric</u> <u>subring</u> $\overline{\mathfrak{A}} \subset \mathfrak{A}$ is defined by

$$\overline{\mathfrak{A}} := \left\{ \sum_{i=0}^{n} \alpha_i Y_i \mid \alpha_i \in \mathbb{Q},\ i = 0,1,\dots,n \right\}.$$

Note that $\overline{\mathfrak{A}}$ is an $(n+1)$-dimensional vector space over $\mathbb{Q}$. To show that $\overline{\mathfrak{A}}$ is indeed a subring of $\mathfrak{A}$ we consider the group G of all linear mappings of $\mathfrak{R}$ into $\mathfrak{R}$ with a matrix DP where D is a diagonal matrix with nonzero elements from $GF(q)$ and P is a permutation matrix. (We call G the <u>monomial</u> <u>group</u>). The order of G is $m := (q-1)^n \cdot n!$. Notice that

$$\forall_{\varphi \in G} \forall_{\underline{v} \in \mathfrak{R}} [w(\varphi(\underline{v})) = w(\underline{v})].$$

If we extend each $\varphi \in G$ to the vector space $\mathfrak{A}$ by defining

$$\varphi\Big(\sum_{\underline{v} \in \mathfrak{R}} a_{\underline{v}} \underline{v} \Big) := \sum_{\underline{v} \in \mathfrak{R}} a_{\underline{v}} \varphi(\underline{v})$$

then

$$\varphi\Big(\sum_{\underline{v} \in \mathfrak{R}} a_{\underline{v}} \underline{v} \Big) * \varphi\Big(\sum_{\underline{w} \in \mathfrak{R}} b_{\underline{w}} \underline{w} \Big) =$$

$$= \sum_{\underline{v} \in \mathfrak{R}} a_{\underline{v}} \varphi(\underline{v}) * \sum_{\underline{w} \in \mathfrak{R}} b_{\underline{w}} \varphi(\underline{w}) =$$

$$= \sum_{\underline{z} \in \mathfrak{R}} \Big(\sum_{\varphi(\underline{v})+\varphi(\underline{w})=\underline{z}} a_{\underline{v}} b_{\underline{w}} \Big) \underline{z} =$$

$$= \varphi\Big(\sum_{\underline{v} \in \mathfrak{R}} a_{\underline{v}} \underline{v} * \sum_{\underline{w} \in \mathfrak{R}} b_{\underline{w}} \underline{w} \Big),$$

i.e. each $\varphi \in G$ is a homomorphism of the ring $(\mathfrak{A}, \oplus, *)$. Since $\mathfrak{A}_\varphi := \{A \in \mathfrak{A} \mid \varphi(A) = A\}$ is a subring it follows that $\bigcap_{\varphi \in G} \mathfrak{A}_\varphi$, which is $\overline{\mathfrak{A}}$, is also a subring of $\mathfrak{A}$.

We now define an <u>averaging-operator</u> T which maps $\mathfrak{A}$ into $\overline{\mathfrak{A}}$ by

$$\forall_{A\in\mathfrak{A}} \left[T(A) := \frac{1}{m} \sum_{\varphi\in G} \varphi(A)\right].$$

Observe that T maps the element $A := \sum_{\underline{v}\in\mathfrak{R}} a_{\underline{v}}\underline{v}$ into $\sum_{i=0}^{n} \alpha_i Y_i$ where α_i is the average of the coefficients $a_{\underline{v}}$ belonging to elements $\underline{v}$ with $w(\underline{v}) = i$. It follows that T acts as the identity on $\overline{\mathfrak{A}}$.

We shall need the following lemma:

(5.4.7) LEMMA: $\forall_{A\in\overline{\mathfrak{A}}}\ \forall_{B\in\mathfrak{A}}\ [T(A*B) = A * T(B)]$.

<u>Proof</u>: $T(A*B) = \frac{1}{m} \sum_{\varphi\in G} \varphi(A*B) = \frac{1}{m} \sum_{\varphi\in G} \varphi(A) * \varphi(B) =$

$$= \frac{1}{m} \sum_{\varphi\in G} A * \varphi(B) = A * T(B).$$

Since

$$S_e = Y_0 \oplus Y_1 \oplus \cdots \oplus Y_e$$

we have by (5.4.5) and (5.4.7)

(5.4.8) $$\mathfrak{R} = T(\mathfrak{R}) = T(S_e * C) = S_e * T(C)$$

and the same holds with C_i instead of C ($i = 1,2,\ldots,e$). Let

(5.4.9) $$\overline{C}_i := T(C_i) \in \overline{\mathfrak{A}} \quad (i = 0,1,\ldots,e).$$

The elements $\overline{C}_i$ are linearly independent elements of the vector space $\overline{\mathfrak{A}}$ over $\mathbb{Q}$ for the same reason that the C_i are linearly independent.

We now define a linear transformation $\mathfrak{S}_e$ of the vector space $\overline{\mathfrak{A}}$ by

(5.4.10) $$\forall_{A\in\overline{\mathfrak{A}}}\ [\mathfrak{S}_e(A) := S_e * A].$$

From (5.4.8) it follows that the e linearly independent elements $\overline{C}_i - \overline{C}_0$ ($i = 1,2, \ldots, e$) of $\overline{\mathfrak{A}}$ satisfy

$$\mathfrak{S}_e(\overline{C}_1 - \overline{C}_0) = 0.$$

Therefore we have proved:

(5.4.11) LEMMA: The linear transformation $\mathfrak{S}_e$ of $\overline{\mathfrak{A}}$ has a kernel of dimension $\geq$ e.

The second part of the proof of Lloyd's theorem uses characters of a group. Although we assume the reader to be familiar with the elements of the theory of group characters we give the definition and a theorem we shall need.

(5.4.12) DEFINITION: A character χ of $\mathfrak{R}$ is a homomorphism of the group $(\mathfrak{R},+)$ into the group $(\mathbb{C}^{\times},)$, i.e. the multiplicative group of $\mathbb{C}$.

(5.4.13) THEOREM: The characters of a finite abelian group $\mathfrak{R}$ form a group isomorphic to $\mathfrak{R}$ (with multiplication as operation).

(5.4.14) DEFINITION: If χ is a non-trivial character on (GF(q),+) and $\underline{v} \in \mathfrak{R}$ we define $\chi_{\underline{v}} : \mathfrak{R} \to \mathbb{C}^{\times}$ by

$$\forall_{\underline{u}\in\mathfrak{R}}\ [\chi_{\underline{v}}(\underline{u}) := \chi((\underline{u},\underline{v}))].$$

Now suppose that $\chi_{\underline{v}} = \chi_{\underline{w}}$, i.e. $\forall_{\underline{u}\in\mathfrak{R}}[\chi((\underline{u},\underline{v})) = \chi((\underline{u},\underline{w}))]$. Then $\forall_{\underline{u}\in\mathfrak{R}}[\chi((\underline{u},\underline{v}-\underline{w}) = 1]$ and since χ is non-trivial on GF(q) we must have $\forall_{\underline{u}\in\mathfrak{R}}[(\underline{u},\underline{v}-\underline{w}) = 0]$, i.e. $\underline{v} = \underline{w}$. Apparently we can define q^n different characters on $\mathfrak{R}$ using the fixed character χ and (5.4.14). We now extend each $\chi_{\underline{v}}$ to a linear functional on $\mathfrak{A}$ (for which we shall use the same symbol) by defining:

(5.4.15)
$$\forall_{A=\sum_{\underline{w}\in\mathfrak{R}} a_{\underline{w}}\underline{w}\in\mathfrak{A}} \left[\chi_{\underline{v}}(A) := \sum_{\underline{w}\in\mathfrak{R}} a_{\underline{w}}\chi_{\underline{v}}(\underline{w})\right].$$

We shall need

(5.4.16) LEMMA: (i) $\forall_{A\in\mathfrak{A}}\forall_{B\in\mathfrak{A}}\ [\chi_{\underline{v}}(A*B) = \chi_{\underline{v}}(A)\chi_{\underline{v}}(B)]$,

$$\text{(ii)}\quad \sum_{\underline{u}\in\mathfrak{R}} \chi_{\underline{v}}(\underline{u})\,\overline{\chi_{\underline{w}}(\underline{u})} = \begin{cases} q^n & \text{if } \underline{v} = \underline{w}, \\ 0 & \text{if } \underline{v} \neq \underline{w}. \end{cases}$$

Proof: (i) $\chi_{\underline{v}}(A*B) = \chi_{\underline{v}}\left(\sum_{\underline{z}\in\mathfrak{R}}\left(\sum_{\underline{u}+\underline{w}=\underline{z}} a_{\underline{u}}b_{\underline{w}}\right)\underline{z}\right) = \sum_{\underline{z}\in\mathfrak{R}}\left(\sum_{\underline{u}+\underline{w}=\underline{z}} a_{\underline{u}}b_{\underline{w}}\right)\chi_{\underline{v}}(\underline{z}) =$

$$= \sum_{\underline{z}\in\mathfrak{R}}\left(\sum_{\underline{u}+\underline{w}=\underline{z}} a_{\underline{u}}b_{\underline{w}}\right)\chi_{\underline{v}}(\underline{u})\,\chi_{\underline{v}}(\underline{w}) = \chi_{\underline{v}}(A)\chi_{\underline{v}}(B)$$

because $\chi_{\underline{v}}(\underline{u}+\underline{w}) = \chi\Big((\underline{v},\underline{u}+\underline{w})\Big) = \chi\Big((\underline{v},\underline{u}) + (\underline{v},\underline{w})\Big) =$

$$= \chi\Big((\underline{v},\underline{u})\Big)\chi\Big((\underline{v},\underline{w})\Big) = \chi_{\underline{v}}(\underline{u})\chi_{\underline{v}}(\underline{w}).$$

(ii) $\sum_{\underline{u}\in\mathfrak{R}} \chi_{\underline{v}}(\underline{u})\,\overline{\chi_{\underline{w}}(\underline{u})} = \sum_{\underline{u}\in\mathfrak{R}} \chi\Big((\underline{u},\underline{v})\Big)\,\overline{\chi\Big((u,w)\Big)} =$

$$= \sum_{\underline{u}\in\mathfrak{R}} \chi_{\underline{u}}(\underline{v}-\underline{w}) = \begin{cases} q^n & \text{if } \underline{v} = \underline{w}\,, \\ q^{n-1}\sum_{\xi\in GF(q)} \chi(\xi) = 0 & \text{if } \underline{v} \neq \underline{w}. \end{cases}$$

From (5.4.16) (ii) it follows that the q^n linear functionals $\chi_{\underline{v}}$ of $\mathfrak{A}$ (where $\underline{v}$ runs through $\mathfrak{R}$) are linearly independent over $\mathbb{C}$. Since there are q^n of them they span the space of all linear functions of $\mathfrak{A}$ (cf. e.g. W. H Greub, Linear Algebra, §2.33).

Consider a mapping $\varphi \in G$ (the monomial group). Then we have

$$\chi_{\varphi(\underline{v})}(Y_k) = \sum_{\underline{w}\in\mathfrak{R},\,w(\underline{w})=k} \chi_{\varphi(\underline{v})}(\underline{w}) = \sum_{\underline{w}\in\mathfrak{R},\,w(\underline{w})=k} \chi\Big((\varphi(\underline{v}),\underline{w})\Big) =$$

$$= \sum_{\underline{w}\in\mathfrak{R},\,w(\underline{w})=k} \chi\Big((\underline{v},\underline{w})\Big) = \chi_{\underline{v}}(Y_k),$$

i.e. the action of the linear functional $\chi_{\underline{v}}$ on $\mathfrak{A}$ is the same for all $\underline{v}$ with the same weight. So if we restrict the linear functionals $\chi_{\underline{v}}$ to $\overline{\mathfrak{A}}$ we find n + 1 linear functionals $\chi_i : \overline{\mathfrak{A}} \to \mathbb{C}^X$ where $\chi_i := \chi_{\underline{v}_i}$ (cf. (5.4.4)). Each linear functional on $\overline{\mathfrak{A}}$ can be extended to a linear functional on $\mathfrak{A}$. Therefore the χ_i $(i = 0,1,\ldots,n)$ must span the space of all linear functionals on $\overline{\mathfrak{A}}$, i.e. they are linearly independent. We can actually explicitly determine the characters χ_i :

(5.4.17) LEMMA: For $k = 0, 1, \ldots, n$ and $w = 0, 1, \ldots, n$ we have

$$\chi_w(Y_k) = \sum_{i=0}^{k} (-1)^i \binom{n-w}{k-i}\binom{w}{i}(q-1)^{k-i} .$$

Proof: (a) From $\chi(0) = 1$ and $\sum_{\alpha \in GF(q)} \chi(\alpha) = 0$ it follows by induction that

$$\sum^{*}_{(\alpha_1,\alpha_2,\ldots,\alpha_i)} \chi(\alpha_1 + \alpha_2 + \cdots + \alpha_i) = (-1)^i,$$

where the * indicates that the sum is taken over all i-tuples of nonzero elements of GF(q).

(b) $$\chi_w(Y_k) = \sum_{\underline{u} \in \mathcal{R}, w(\underline{u})=k} \chi\big((\underline{v}_w, \underline{u})\big) =$$

$$= \sum_{i=0}^{k} \binom{w}{i}\binom{n-w}{k-i}(q-1)^{k-i} \sum^{*}_{(\alpha_1,\alpha_2,\ldots,\alpha_i)} \chi(\alpha_1+\alpha_2+ \cdots + \alpha_i)$$

and the result follows from (a).

(5.4.18) COROLLARY: $$\chi_w(S_e) = \sum_{i=0}^{e} (-1)^i \binom{w-1}{i}\binom{n-w}{e-i}(q-1)^{e-i} .$$

Proof: From (5.4.17) we find

$$\chi_w(S_e) = \sum_{k=0}^{e}\sum_{i=0}^{k} (-1)^i \binom{w}{i}\binom{n-w}{k-i}(q-1)^{k-i} =$$

$$= \sum_{i=0}^{e} (-1)^i \binom{w}{i} \sum_{k=i}^{e} \binom{n-w}{k-i}(q-1)^{k-i} =$$

$$= \sum_{i=0}^{e} (-1)^i \binom{w}{i} \sum_{\ell=0}^{e-i} \binom{n-w}{\ell}(q-1)^{\ell}.$$

If we write $\binom{w}{i} = \binom{w-1}{i-1} + \binom{w-1}{i}$ and rearrange the sum we obtain the required result.

This completes the second part of the preliminaries and we can now formulate and prove Lloyd's theorem.

(5.4.19) THEOREM: If a perfect e-error-correcting code of block length n over GF(q) exists then the polynomial

$$P_e(x) := \sum_{i=0}^{e} (-1)^i \binom{n-x}{e-i}\binom{x-1}{i}(q-1)^{e-i} ,$$

where $\binom{a}{i} := a(a-1) \cdots (a-i+1) / i!$,

has e distinct integral zeros among 1, 2, ..., n-1.

Proof: Consider the linear transformation $\mathbf{S}_e$ of $\mathfrak{A}$ defined in (5.4.10) and let K be the kernel of $\mathbf{S}_e$. In (5.4.11) it was shown that dim (K) $\geq$ e. By definition $S_e * A = 0$ if $A \in K$ and then it follows from (5.4.16)(i) that for w = 0, 1, ..., n and all $A \in K$

$$\chi_w(S_e)\chi_w(A) = 0.$$

Suppose r of the characters vanish identically on K. Apparently $\chi_w(S_e) = 0$ for the remaining n + 1 - r characters. Since the characters χ_w were linearly independent we must have dim (K) $\leq$ n + 1 - r. In (5.4.18) we found that $P_e(w) = \chi_w(S_e)$. For at least e values of w we must have $P_e(w) = 0$ and since $P_e(x)$ is a polynomial of degree e we see that all the zeros of $P_e(x)$ are integers in [0,n]. By substitution we find

$$P_e(0) = \sum_{i=0}^{e} \binom{n}{e-i}(q-1)^{e-i} \neq 0,$$

$$P_e(n) = (-1)^e\binom{n-1}{e} \neq 0 \text{ if } n > e.$$

This proves the theorem.

The following theorem gives us some idea of the interpretation of the zeros of $P_e(x)$.

(5.4.20) THEOREM: If $\underline{v} \neq \underline{0}$ is a vector in the dual of a linear perfect e-error-correcting code then $w(\underline{v})$ is a zero of the polynomial $P_e(x)$.

Proof: If $\underline{v}$ is in the dual of the perfect e-error-correcting code C and $\underline{v} \neq \underline{0}$ then by (5.4.14) and (5.4.15)

$$\chi_{\underline{v}}(C) = \sum_{\underline{c} \in C} \chi\big((\underline{c},\underline{v})\big) = \sum_{\underline{c} \in C} \chi(0) = q^k$$

if k is the dimension of C. By (5.4.16) (i) and (5.4.3) we have

$$\chi_{\underline{v}}(S_e)\chi_{\underline{v}}(C) = \chi_{\underline{v}}(\mathcal{R}^{(n)}) = 0.$$

Hence

$$P_e\big(w(\underline{v})\big) = \chi_{w(\underline{v})}(S_e) = \chi_{\underline{v}}(S_e) = 0.$$

For future use we put $P_e(x)$ in an other form. To do this we apply the same reduction that occurred in (5.2.3). First we remark

$$\sum_{i=0}^{e-j} \binom{n-x-j}{e-i-j}\binom{x-1}{i} = \binom{n-j-1}{e-j}.$$

This simply follows from multiplying the Taylor series of $(1+z)^{n-x-j}$ and $(1+z)^{x-1}$ and considering the coefficient of z^{e-j}. Then we have

$$P_e(x) = \sum_{i=0}^{e} (-1)^i \binom{n-x}{e-i}\binom{x-1}{i} \sum_{j=0}^{e-i} \binom{e-i}{j} q^j (-1)^{e-i-j} =$$

$$= (-1)^e \sum_{j=0}^{e} (-1)^j q^j \sum_{i=0}^{e-j} \binom{n-x}{e-i}\binom{e-i}{j}\binom{x-1}{i} =$$

$$= (-1)^e \sum_{j=0}^{e} (-1)^j q^j \binom{n-x}{j} \sum_{i=0}^{e-j} \binom{n-x-j}{e-i-j}\binom{x-1}{i}, \quad \text{i.e.}$$

(5.4.21) $$P_e(x) = (-1)^e \sum_{j=0}^{e} (-1)^j q^j \binom{n-x}{j}\binom{n-j-1}{e-j}.$$

5.5 Nonexistence theorems

We shall now combine (5.4.19) (or (5.4.21)) and (5.2.2) (or (5.2.3)) to show the nonexistence of certain perfect codes. We shall consider codes with one word and repetition codes (2 words) as trivial perfect codes.

In the following we assume the existence of a perfect e-error-correcting code $(1 < e < n)$ over $GF(q)$, where $q = p^{\alpha}$, and we shall try to arrive at a contradiction. By (5.4.19) there are e distinct integers $1 < x_1 < x_2 < \cdots < x_e < n$ which are the zeros of the polynomial $P_e(x)$. By substitution in (5.4.19) we find

$$P_e(1) = \binom{n-1}{e}(q-1)^e \neq 0 \quad \text{since} \quad e \leq n-1$$

and

$$P_e(2) = \frac{1}{e}\binom{n-2}{e-1}(q-1)^{e-1}\{q(n-e-1) - (n-1)\} = 0 \quad \text{only if}$$

$q = 1 + \frac{e}{n-e-1}$ which implies $e \geq \frac{n-1}{2}$. Since the minimum distance of the code is $2e + 1 \leq n$ we must have $e = \frac{n-1}{2}$, $q = 2$ and then we have the parameters of a repetition code. Therefore we may assume that $x_1 > 2$.

The following lemma will be the basis of our theorems:

(5.5.1) LEMMA: The zeros of $P_e(x)$ satisfy the relations:

$$\text{(i)} \quad x_1 + x_2 + \cdots + x_e = \frac{e(n-e)(q-1)}{q} + \frac{e(e+1)}{2} ,$$

$$\text{(ii)} \quad x_1 x_2 \cdots x_e = e!\, q^{k-e}$$

(where k is the exponent in the right-hand side of (5.2.2)).

Proof: We saw in the proof of (5.4.19) that

$$P_e(0) = \sum_{i=0}^{e} \binom{n}{i}(q-1)^i = q^k \quad \text{by (5.2.2).}$$

From (5.4.21) we see that the coefficient of x^e in $P_e(x)$ is $(-1)^e q^e/e!$.

Also from (5.4.21) we find the coefficient of x^{e-1} in $P_e(x)$ to be

$$-(-1)^e \frac{q^e}{e!} \sum_{i=0}^{e-1} (n-i) + \frac{(-1)^e q^{e-1}}{(e-1)!}(n-e) = \frac{(-1)^{e+1} q^e}{e!} \left\{\frac{e(2n-e+1)}{2} - \frac{e(n-e)}{q}\right\} .$$

From these coefficients we find the sum and the product of the zeros of $P_e(x)$.

(5.5.2) COROLLARY: If a perfect e-error-correcting code of block length n over GF(q) exists then

$$e(n-e) \equiv 0 \pmod{q}.$$

Let us now consider perfect 2-error-correcting codes. From (5.4.21) and (5.2.2) we find

$$2P_2(x) = (qx)^2 - \{(2n-1)q - (2n-4)\}(qx) + 2q^k.$$

If $q = 2$ we know from Section 5.2 that $(2n+1)^2 + 7 = 2^{k+3} = 8P_2(0)$. In this case Lloyd's theorem says that the equation

$$y^2 - 2(n+1)y + 2^{k+1} = 0$$

has two distinct roots y_1 and y_2 which are even integers. We saw above that we may assume that y_1 and y_2 are both > 4. Since $y_1y_2 = 2^{k+1}$ we must have $y_1 = 2^a$, $y_2 = 2^b$ where $3 \leq a < b$. Then $2(n+1) = 2^a + 2^b$ and the equation $(2n+1)^2 + 7 = 2^{k+3}$ becomes

$$(2^a+2^b-1)^2 = 2^{a+b+2} - 7$$

and since a and b are both ≥ 3 this implies $1 \equiv 7 \pmod{16}$ which is a contradiction. This then proves Theorem (5.2.11). Subsequently we look at $e = 2$, $q > 2$. The equation (5.2.2) is a quadratic equation in n which we can solve, expressing n in q and k. For the roots of the equation $P_2(x) = 0$ we then find from (5.5.1)

(5.5.3)
$$x_1x_2 = 2q^{k-2},$$
$$q(x_1+x_2) = 1 + (8q^k+q^2-6q+1)^{1/2}.$$

Since we may assume that x_1 and x_2 are > 2 we see that one of the roots is p^λ and the other $2p^\mu$ where $\lambda \geq 1$, $\mu \geq 1$, $\lambda + \mu = \alpha(k-2)$. Substitution in (5.5.3) and elimination of the square root yields

(5.5.4)
$$8q^{k-1} + q - 6 = q(p^\lambda+2p^\mu)^2 - 2(p^\lambda+2p^\mu).$$

Since λ and μ are positive all the terms on the two sides of (5.5.4) are divisible by p, i.e. $p = 2$ or $p = 3$. If $p = 2$ then the right-hand side of (5.5.4) is divisible by 4 whereas this holds for the left-hand side only if $q = 2$. The case $e = 2$, $q = 2$

was settled above. It remains to consider $p = 3$. First assume $q = 3$. Then (5.5.4) reduces to

$$8\cdot 3^{k-1} - 3 = 3(3^{\lambda} + 2\cdot 3^{\mu})^2 - 2(3^{\lambda} + 2\cdot 3^{\mu})$$

which implies $\mu = 1$, $\lambda = 2$, i.e. $k = 5$ and hence $n = 11$ (from (5.2.2)). If, on the other hand, $q = 3^{\alpha}$ with $\alpha > 1$ then (5.5.4) implies

$$3 \equiv 3^{\lambda} + 2\cdot 3^{\mu} \pmod 9.$$

From this we find $\lambda = 1$, $\mu > 1$ and then (5.5.4) reduces to

$$q^{k-1} = 2q + q\cdot 3^{2\mu} - 3^{\mu}$$

which is impossible because the left-hand side is clearly divisible by a higher power of 3 than the right-hand side. Now all possibilities have been discussed and we have:

(5.5.5) <u>THEOREM</u>: <u>The only non-trivial perfect</u> 2-<u>error-correcting code over any alphabet</u> GF(q) <u>is the ternary Golay code</u>.

In order to be able to use Lloyd's theorem for $e \geq 3$, where explicitly solving $P_e(x) = 0$ is either impossible or too much work, we take a closer look at the polynomial $P_e(x)$. Notice that if a is any positive integer then $\binom{a}{i} = \frac{a(a-1)\cdots(a-i+1)}{i!} \geq 0$ because a negative factor can occur in the numerator only if some other factor is 0. Since we are interested in the value of $P_e(x)$ for integral x the sum in (5.4.19) is an alternating sum (if $1 \leq x \leq n-1$). The terms in this sum decrease in absolute value if $x < \frac{(n-e+1)(q-1)+e}{(q-1)+e}$. Therefore $P_e(x)$ cannot be 0 for such integral values of x. This proves

(5.5.6) <u>LEMMA</u>: If $P_e(x)$ has e integral zeros in $[1,n-1]$ then

$$x_1 \geq \frac{(n-e+1)(q-1)+e}{(q-1)+e} .$$

<u>Remark</u>: In the same way that (5.5.6) was proved we can find a lower bound for x_1 from (5.4.4) but this bound is not as good as (5.5.6).

From (5.4.21) we can easily find an estimate for q. As we remarked in the proof of (5.4.19) we have

$$P_e(n) = (-1)^e \binom{n-1}{e} \neq 0$$

From (5.4.21) we find

$$P_e(n-1) = (-1)^2 \left\{\binom{n-1}{e} - q\binom{n-2}{e-1}\right\} = (-1)^e \binom{n-1}{e}\{1 - \frac{qe}{n-1}\}$$

and by (5.4.19) this must have the same sign as $P_e(n)$, i.e.

(5.5.7) THEOREM: <u>If a perfect</u> e-<u>error-correcting code of block length</u> n <u>over</u> GF(q) <u>exists</u> ($e < n$) <u>then</u>

$$q \leq (n-1)/e.$$

We are now in a position to prove a first nonexistence theorem.

(5.5.8) THEOREM: <u>If</u> $e \geq 3$, $q = p^\alpha$ <u>with</u> $p > e$ <u>then there is no non-trivial perfect</u> e-<u>error-correcting code over</u> GF(q).

Proof: (i) A typical term in (5.2.3) has the form

(5.5.9) $$t_j := (-1)^j q^j \frac{n(n-1)\cdots(n-j+1)}{j!} \frac{(n-j-1)(n-j-2)\cdots(n-e)}{(e-j)!} .$$

Since $p > e$ we find from (5.5.2) that $q \mid (n-e)$. In the fractions in (5.5.9) the only term in the numerator which is divisible by p is $n - e$. No term in the denominator is divisible by p. Let $p^\sigma \| (n-e)$. Then $p^{\alpha j+\sigma} \| t_j$ for $j = 0, 1, \ldots, e-1$ and $p^{\alpha e} \| t_e$. By (5.2.3) the expression $\sum_{j=0}^{e} t_j$ is divisible by $p^{\alpha k}$ where $k > e$. This is only possible if the two terms t_j containing the lowest powers of p are divisible by the <u>same</u> power of p. Hence $\alpha e = \sigma$. Therefore we have proved that $q^e \mid (n-e)$.

(ii) From (i) it follows that the first term on the right-hand side of (5.5.1) (i) is divisible by q^{e-1}. The term $\frac{e(e+1)}{2}$ is divisible by p only if $p = e+1$. This implies that at least one zero of $P_e(x)$ is not divisible by p unless $p = e+1$ in which case we know that at least one zero of $P_e(x)$ is not divisible by p^2. It then follows from (5.5.1) (ii) that at least one of the zeros of $P_e(x)$ is a divisor of $(e+1)!$ Hence $x_1 \leq (e+1)!$

(iii) Since $q > e$ and $q^e | (n-e)$ we have $(n-e) \geq (e+1)^e$. Then Lemma (5.5.6) yields $x_1 > 1 + \frac{1}{2}(e+1)^e$. Combined with (ii) we find $1 + \frac{1}{2}(e+1)^e \leq (e+1)!$ which is false for $e \geq 3$. This proves the theorem.

Remark: The argument in the proof of (5.5.8) goes through for $e = 2$. In that case we do not arrive at a contradiction but we find that $x_1 = 6$, $q = 3$ is the only possibility. This again leads to the ternary Golay code.

We now refine the argument used in the proof of (5.5.8) to prove:

(5.5.10) THEOREM: *If* $e \geq 3$, $q = p^\alpha > e$ *and* $p < e$, $p \nmid e$ *then there is no non-trivial perfect* e*-error-correcting code over* GF(q).

Proof: (i) As in (5.5.8) we find $q | (n-e)$ from (5.5.2). From (5.5.9) we find

$$t_{j+1}/t_j = -q\,\frac{(n-j)(e-j)}{(j+1)(n-j-1)} \qquad (0 \leq j \leq e-1).$$

Now, since $p \nmid e$ and $p \mid (n-e)$ we have $p \nmid n$ which implies that at most one of the factors in the denominator is divisible by p. Furthermore, with the exception of $j = e-1$, the highest power of p which divides the denominator is less than q because $q | (n-e)$ and $q > e$. Therefore the terms in (5.5.9) containing the lowest powers of p are t_0 and t_e. Again, let $p^\sigma \| (n-e)$. Since $q | (n-e)$ and $q > e$ the corresponding terms in the numerator and the denominator of

$$\frac{(n-1)(n-2)\cdots(n-e+1)}{(e-1)(e-2)\cdots\quad 1}$$

are divisible by the same power of p. Hence $p^\sigma \| \binom{n-1}{e} = t_0$. In the same way we show that $p^{\alpha e} \| t_e$. So again we find $\sigma = \alpha e$, i.e. $q^e | (n-e)$.

Part (ii) and (iii) of the proof of (5.5.8) can now be copied to complete the proof of (5.5.10).

For primes p which divide e the result which we can obtain is weaker:

(5.5.11) **THEOREM**: *Let* $p | e$, $q = p^\alpha$ *and let* $q > e$ *if* $p > 2$ *and* $q > 2e$ *if* $p = 2$. *Define*

$$M_p(e) := \begin{cases} 2e! + e - 1 & \text{if } p > 2, \\ \left((e-1)!\right)_1 e + e - 1 & \text{if } p = 2, \end{cases}$$

where $(a)_1$ *denotes the largest odd factor of* a. *If a non-trivial perfect* e-*error-correcting code of block length* n *over* GF(q) *exists, then* $n < M_p(e)$.

Proof: (i) Suppose all the zeros of $P_e(x)$ are divisible by a higher power of p than is contained in $\frac{1}{2} e(e+1)$. Then in (5.5.1) (i) the two terms on the right-hand side are divisible by the same power of p and therefore $p^\alpha \| 2(n-e)$. Assume $p > 2$. Then, just as in the proof of (5.5.10), we see that numerator and denominator of

$$\frac{(n-1)(n-2) \cdots (n-e+1)}{(e-1)(e-2) \cdots \quad 1}$$

are divisible by the same power of p and hence $q \nmid \binom{n-1}{e}$ contradicting (5.2.3). If $p = 2$ the same reasoning holds because then $p^{\alpha-1} \| (n-e)$ and we required that $p^{\alpha-1} = \frac{1}{2} q > e$.

(ii) Since we found a contradiction in (i) the assumption was false. Hence there is a zero of $P_e(x)$ which is not divisible by a higher power of p than $\frac{1}{2} e(e+1)$. If $p > 2$ this implies $x_1 \leq e!$ and if $p = 2$ it implies $x_1 \leq \left((e-1)!\right)_1 \cdot (\frac{1}{2} e)$. The theorem then follows from (5.5.6).

The theorems and methods of (5.5.8) to (5.5.11) are sufficient to solve the problem of the existence of perfect e-error-correcting codes, given e. This has been done for $e \leq 7$ yielding no new perfect codes (cf. [9], [10], [11]). We shall demonstrate the method for $e = 3$. If a perfect 3-error- correcting code of block length n over GF(q) exists ($q = p^\alpha$) then by (5.5.8) $p \leq 3$. By (5.5.10) $p = 2$

implies $q = 2$ which case was completely treated in (5.2.10). If $p = 3$ then (5.5.11) implies $q = 3$ or $q > 3$ and $n < 14$. In the latter case (5.5.7) implies $q < 5$ and this is a contradiction. Apparently only $q = 3$ has to be considered. If $q = 3$ we find from (5.5.1) and (5.5.6):

$$x_1 + x_2 + x_3 = 2n,$$
$$x_1 x_2 x_3 = 2 \cdot 3^{k-2},$$
$$x_1 \geq \frac{2n - 1}{5}.$$

This is impossible because the second equation shows that the three zeros have the form $2 \cdot 3^a$, 3^b, 3^c. Since these zeros are distinct the sum of the zeros is $\geq 7x_1 > 2n$ contradicting the first equation. We have proved:

(5.5.12) THEOREM: The only non-trivial perfect 3-error-correcting code over any alphabet GF(q) is the binary Golay code.

It seems that only very little more than the material presented in this section should be enough to completely settle the question of the existence of non-trivial perfect codes.

Since the appearance of the first printing of these lecture notes the question has indeed been settled by A. Tietäväinen (cf. [19]) and independently by W.A. Zinoviev and W.K. Leontiev (cf. [20]). The proofs depend on a refinement of the arithmetic-geometric mean inequality which is applied to (i) and (ii) of (5.5.1). It has also been shown by H.W. Lenstra jr. (cf. [21]) that Theorem (5.4.19) is also true if q, the number of symbols in the alphabet, is not a power of a prime. The question of the existence of perfect codes in this case remains open. A survey perfect codes and connected material can be found in [22] .

VI. WEIGHT ENUMERATION

6.1 The MacWilliams equations

There is a remarkable relation between the weight enumerator of a linear code and the weight enumerator of the dual code. The relation was first discovered by F. J. MacWilliams. The proof we give here is based on an idea due to A. M. Gleason.

We shall first prove a lemma which resembles the Möbius inversion formula.

(6.1.1) **LEMMA:** Let χ be a non-trivial character on $(GF(q),+)$ and let $\chi_{\underline{v}}: \mathcal{R} \to \mathbb{C}^+$ be defined as in (5.4.14). If A is a vector space over $\mathbb{C}$, $f: \mathcal{R} \to A$ and if $g: \mathcal{R} \to A$ is defined by

$$\forall_{\underline{u}\in\mathcal{R}} \left[g(\underline{u}) := \sum_{\underline{v}\in\mathcal{R}} f(\underline{v})\chi_{\underline{v}}(\underline{u}) \right]$$

then for any linear subspace $V \subset \mathcal{R}$ and the dual space $V^{\perp}$ we have

$$\frac{1}{|V|} \sum_{\underline{u}\in V} g(\underline{u}) = \sum_{\underline{v}\in V^{\perp}} f(\underline{v}).$$

Proof: $$\sum_{\underline{u}\in V} g(\underline{u}) = \sum_{\underline{u}\in V}\sum_{\underline{v}\in\mathcal{R}} f(\underline{v})\chi_{\underline{v}}(\underline{u}) = \sum_{\underline{v}\in\mathcal{R}} f(\underline{v}) \sum_{\underline{u}\in V} \chi((\underline{u},\underline{v})) =$$

$$= |V| \sum_{\underline{v}\in V^{\perp}} f(\underline{v}) + \sum_{\underline{v}\notin V^{\perp}} f(\underline{v}) \sum_{\underline{u}\in V} \chi((\underline{u},\underline{v})).$$

In the inner sum of the second term $(\underline{u},\underline{v})$ takes on every value $\in GF(q)$ the same number of times and since $\sum_{\alpha\in GF(q)} \chi(\alpha) = 0$ for every non-trivial character this proves the theorem.

We apply this theorem with $A :=$ space of polynomials in two variables ξ and η with coefficients in $\mathbb{C}$, $f(\underline{v}) := \xi^{w(\underline{v})}\, \eta^{n-w(\underline{v})}$. Then we find (using $w(a) := 1$ if $a \neq 0$, $w(0) = 0$):

$$g(\underline{u}) = \sum_{v_1 \in GF(q)} \cdots \sum_{v_n \in GF(q)} \xi^{w(v_1)+\cdots+w(v_n)} \eta^{(1-w(v_1))+\cdots+(1-w(v_n))} \chi(u_1 v_1 + \cdots + u_n v_n) =$$

$$= \prod_{i=1}^{n} \Big(\sum_{v \in GF(q)} \xi^{w(v)} \eta^{1-w(v)} \chi(u_i v) \Big) .$$

Since the inner sum is $\eta + (q-1)\xi$ if $u_i = 0$ and $\eta + \xi \Big[\sum_{\substack{\alpha \in GF(q) \\ \alpha \neq 0}} \chi(\alpha) \Big] = \eta - \xi$ if $u_i \neq 0$ we find

$$g(\underline{u}) = \big(\eta + (q-1)\xi\big)^{n-w(\underline{u})} (\eta - \xi)^{w(\underline{u})} .$$

Now let V be a linear (n,k) code over GF(q) and $V^{\perp}$ the dual code. Let $A(z) := \sum_{i=0}^{n} A_i z^i$ and $B(z) := \sum_{i=0}^{n} B_i z^i$ be the weight enumerators of V and $V^{\perp}$.

Apply (6.1.1) with $f(\underline{v}) = \xi^{w(\underline{v})} \eta^{n-w(\underline{v})}$ where $\xi = z$, $\eta = 1$. We find

$$q^{-k}\big(1 + (q-1)z\big)^n A\Big(\frac{1 - z}{1 + (q-1)z} \Big) = B(z).$$

This is the MacWilliams identity i.e.

(6.1.2) <u>THEOREM</u>: <u>If</u> A(z) <u>is the weight enumerator of the linear</u> (n,k) <u>code over</u> GF(q) <u>and</u> B(z) <u>is the weight enumerator of the dual code then</u>

$$q^{-k}\big(1 + (q-1)z\big)^n A\Big(\frac{1 - z}{1 + (q-1)z} \Big) = B(z).$$

<u>Example</u>: By (2.3.14) the weight enumerator of the first order RM code of length 2^m is

$$A(z) = 1 + (2^{m+1} - 2)z^{2^{m-1}} + z^{2^m} .$$

Hence

$$B(z) = 2^{-m-1}(1 + z)^{2^m} A\Big(\frac{1 - z}{1 + z}\Big)$$

is the weight enumerator of the extended Hamming code of length 2^m (by (2.3.7) and (2.3.8)). (See problem (2.6.6)).

The formula (6.1.2) can be very useful in weight enumeration problems. We shall illustrate this for $q = 2$ to simplify the calculations but everything goes through analogously for $q > 2$. Suppose that the coefficients A_i of the weight enumerator $\sum_{j=0}^{n} A_j z^j$ of an (n,k) code are known except for s values of the index j. If $B_0, B_1, \ldots, B_{s-1}$ are known, which for instance is the case if the dual code has minimum weight s ($B_0 = 1$, $B_1 = B_2 = \cdots = B_{s-1} = 0$), then we can determine the unknown A_j's as follows. From (6.1.2) we have

$$\sum_{j=0}^{n} A_j z^j = 2^{k-n} \sum_{j=0}^{n} B_j (1-z)^j (1+z)^{n-j} .$$

Differentiate both sides ℓ times, where $0 \leq \ell < s$, divide by $\ell!$ and substitute $z = 1$. We find:

$$\text{(6.1.3)} \qquad \sum_{j=0}^{n} \binom{j}{\ell} A_j = 2^{k-\ell} \sum_{j=0}^{\ell} (-1)^j \binom{n-j}{\ell-j} B_j \qquad (\ell = 0,1,\ldots,s-1)$$

which is a system of s linear equations in the unknowns $A_{j_1}, A_{j_2}, \ldots, A_{j_s}$. The coefficient matrix of this system is

$$[a_{\ell k}) := \left[\binom{j_k}{\ell} \right], \qquad (\ell = 0,1,\ldots,s-1;\ k = 1,2,\ldots,s).$$

Since $[a_{\ell k}]$ can be transformed into the Vandermonde matrix by elementary row operations the equations are independent.

6.2 Weight enumeration of RM codes

In the past few years much work has been done on the weights in RM codes. In this section we shall discuss some of the results briefly.

First we consider a theorem due to G. Solomon and R. J. McEliece.

(6.2.1) THEOREM: *Let $h(x)$ be the check polynomial of a binary cyclic code of length n (odd). If $h(x)$ does not have a pair of zeros with product 1 then the weight of every code word is divisible by 4.*

Proof: Let $g(x)$ be the generator of the code. Since $1.1 = 1$ the factor $x - 1$ divides $g(x)$ and not $h(x)$. Therefore every code word has even weight. If $c(x) = \sum_{i=1}^{d} x^{e_i}$ is a code word of weight d then the condition on $h(x)$ implies that $c(x)c(x^{-1}) = 0$ (in the ring $\mathcal{R}$). Now by the same argument that was used to prove (4.4.5) we must have $d(d-1) \equiv 0 \pmod 4$, i.e. $d \equiv 0 \pmod 4$.

Remark: Theorem (6.2.1) was generalized by R. J. McEliece (On periodic sequences from GF(q), Journal of Comb. Theory 10 (1971), 80-91). In the generalized form "pair" is replaced by "j-tuple" and 4 is replaced by 2^j. The proof is much more difficult. A consequence of this theorem is that all the weights in the r-th order RM code of length 2^m are divisible by $2^{[(m-1)/r]}$. (cf. (4.3.2) and (4.3.5); also see (6.2.3).

For some time it was conjectured that a primitive BCH-code of designed distance d has minimum distance d. As an application of (6.2.1) we show that this is false. Let $g(x)$ be the generator of the BCH code of length 127 and designed distance $d = 29$. Then the words of even weight form a subcode with generator $(x-1)g(x)$ and check polynomial $h(x)$. Here $h(x)$ has degree 42 and it is easily checked that $h(x)$ satisfies the conditions of (6.2.1). Therefore the minimum weight in this subcode is divisible by 4. We now annex the all-one vector to the generator matrix of the subcode to find the original code back again. Since we know from (4.1.11) that the minimum distance of the BCH code is odd this minimum distance must be $\equiv 3 \pmod 4$. Therefore the minimum distance is not 29. In fact it is 31.

The result on weights in RM codes mentioned in the remark above can be proved in different ways. One of them depends on the following theorem:

(6.2.2) THEOREM: Let $\mathcal{R}^{(m)}$ be the m-dimensional vector space over GF(2) and let $F: \mathcal{R}^{(m)} \to GF(2)$ be a polynomial, $F = F(x_1, x_2, \ldots, x_m)$, of degree r. We shall say $G \subset F$ if the monomials of G form a subset of the monomials of F. We define $\nu(G)$ to be the number of variables not involved in G and we let $|G|$ denote the number of monomials of G.

<u>If</u> $N(F)$ <u>is the number of zeros of</u> F <u>in</u> $\mathcal{R}^{(m)}$ <u>then</u>

$$N(F) = 2^{m-1} + \sum_{G\subset F} (-1)^{|G|} 2^{|G|+\nu(G)-1} .$$

<u>Proof</u>: For every $G \subset F$ define $f(G)$ to be the number of points in $\mathcal{R}^{(m)}$ where all the monomials of G have the value 0 and all the other monomials of F have the value 1. Clearly we have

$$\sum_{H\subset G} f(H) = 2^{\nu(F-G)}$$

(because this is the number of points in the affine subspace of $\mathcal{R}^{(m)}$ defined by $x_{i_1} = x_{i_2} = \cdots = x_{i_\nu} = 1$ where the x_{i_k} are the variables occurring in $F - G$.) It follows from (4.2.7) that

$$f(G) = \sum_{H\subset G} \mu(H,G) 2^{\nu(F-H)}.$$

Clearly

$$N(F) = \sum_{G\subset F,\ |F-G|\equiv 0 \pmod 2} f(G).$$

Since $\sum_{G\subset F} f(G) = 2^m$ we find

$$\begin{aligned} N(F) &= 2^{m-1} + \frac{1}{2}\sum_{G\subset F} (-1)^{|F-G|} f(G) = \\ &= 2^{m-1} + \frac{1}{2}\sum_{G\subset F} (-1)^{|F-G|} \sum_{H\subset G} \mu(H,G) 2^{\nu(F-H)} = \\ &= 2^{m-1} + \frac{1}{2}\sum_{H\subset F} (-1)^{|F-H|} 2^{\nu(F-H)} \sum_{H\subset G\subset F} 1 = \\ &= 2^{m-1} + \frac{1}{2}\sum_{H\subset F} (-1)^{|F-H|} 2^{\nu(F-H)} 2^{|F-H|} = \\ &= 2^{m-1} + \frac{1}{2}\sum_{G\subset F} (-1)^{|G|} 2^{\nu(G)+|G|} . \end{aligned}$$

(Remark: This proof is new. The theorem is due to R. J. McEliece).

We now apply this theorem to RM codes. Following the description on page 29 the r-th order RM code is the set of value-tables of polynomials of degree $\leq r$ in m variables $x_1, x_2, \ldots, x_m$ over GF(2). The code word corresponding to the polynomial F has weight $2^m - N(F)$. If $G \subset F$ and G has degree d then $\nu(G) \geq m - |G|d$, i.e. $|G| \geq \lceil \frac{m - \nu(G)}{d} \rceil$ (where we use $\lceil a \rceil := \min\{n \in \mathbb{Z} | n \geq a\}$). Since

$$\nu(G) + \left\lceil \frac{m - \nu(G)}{d} \right\rceil \geq \lceil \tfrac{m}{d} \rceil$$

we have proved (by applying (6.2.2):

(6.2.3) <u>THEOREM</u>: <u>The weights of the code words in the</u> r-th <u>order</u> RM <u>code of length</u> 2^m <u>are divisible by</u> $2^{\lceil \frac{m}{r} \rceil - 1}$.

We now turn to the problem of weight enumeration for 2-nd order RM codes. For this purpose we must study quadratic forms in m variables. First, we omit the all-one vector from the basis of the RM code. Then, since $x_i^2 = x_i$, all code words are value-tables of homogeneous quadratic forms $\sum_{i \leq j} a_{ij} x_i x_j$. An invertible linear transformation of $\mathcal{R}^{(m)}$, i.e. a change of variables to say $y_1, y_2, \ldots, y_m$ maps all quadratic forms in $x_1, \ldots, x_m$ into quadratic forms in $y_1, \ldots, y_m$. Since the weight of a code word is $2^m - N(F)$ if F is the corresponding quadratic form, the linear transformation does not change the weights. Therefore the following lemma will be useful. We consider nonsingular quadratic forms, i.e. forms F with $\nu(F) = 0$.

(6.2.4) <u>LEMMA</u>: A non-singular homogeneous quadratic form $F(x_1, \ldots, x_m) = \sum_{i \leq j} a_{ij} x_i x_j$ can be transformed by an invertible linear transformation into a quadratic form of the type

$$x_1 x_2 + Q(x_3, \ldots, x_m).$$

<u>Proof</u>: We know from (2.3.5) that F is not identically 1 or 0. If $F(\xi_1, \ldots, \xi_m) = 0$ and $(\xi_1, \ldots, \xi_m) \neq \underline{0}$ then we can find a transformation of type

$$x_i = \xi_i x_1' + \cdots \qquad (i = 1, \ldots, m)$$

which transforms F into a quadratic form $F'(x_1',\dots,x_n') = \sum_{i\le j} a_{ij}' x_i' x_j'$ in which $a_{11}' = 0$. Without loss of generality we may assume $a_{12}' = 1$. A transformation with $x_2'' = \sum_{i=2}^{m} a_{1i}' x_i'$ then yields a quadratic form

$$F''(x_1'',\dots,x_m'') = x_1''x_2'' + \sum_{2\le i\le j} a_{ij}''\, x_i''\, x_j'' .$$

If all the a_{2j}'' are 0 we are finished. Otherwise let $y_1 := x_1'' + \sum_{2\le j} a_{2j}'' x_j''$, $y_i = x_i''$ for $i \ge 2$. We then obtain the desired form.

With the aid of this lemma we can prove the following theorem due to T. Kasami:

(6.2.5) **THEOREM**: The weight of every code word in the second order RM code of length 2^m is of the form

$$2^{m-1} + \epsilon\, 2^{\ell}$$

where $\epsilon = 0$ or ± 1 and $\frac{m}{2} - 1 \le \ell \le m - 1$.

Proof: First observe that we can remove the all-one vector from the basis without changing the assertion. By the definition on page 29 (see the example on page 29) every basis vector $\underline{v}_i$, $i = 1, \dots, m-1$ in the basis of the RM codes of length 2^m is of the form $(\underline{v}_i', \underline{v}_i')$ where $\underline{v}_i'$ is a basis vector for the RM codes of length 2^{m-1}. In the same way we see that for $i = 1, 2, \dots, m-2$ every basis vector $\underline{v}_i$ has the form $(\underline{v}_i'',\underline{v}_i'',\underline{v}_i'',\underline{v}_i'')$ where $\underline{v}_i''$ is a basis vector for the RM codes of length 2^{m-2}. This enables us to prove the theorem by induction. For small values of m it is easy to check the theorem. If $F(x_1,\dots,x_m)$ is a singular quadratic form we may assume without loss of generality that x_m does not occur and then the weight of the corresponding code word is twice the weight of a code word in the second order RM code of length 2^{m-1}. If F is non-singular we apply (6.2.4) to transform F into $x_{m-1}x_m + Q(x_1,\dots,x_{m-2})$. Now $\underline{v}_{m-1} = (\underline{0}, \underline{1}, \underline{0}, \underline{1})$ and $\underline{v}_m = (\underline{0}, \underline{0}, \underline{1}, \underline{1})$ where $\underline{0}$ and $\underline{1}$ are of length 2^{m-2}. It follows that the weight of the code word corresponding to F is $3d + (2^{m-2} - d)$ where d is the weight of a code word in the second order RM code of length 2^{m-2}. So in both cases the assertion follows from the induction hypothesis.

(Remark: This proof is new. Kasami's proof depended on a standard form of quadratic functions derived by extending (6.2.4).)

Combining the results of this section and section 6.1 and by generalizing (6.2.4) N. J. A. Sloane and E. R. Berlekamp have determined the complete weight enumerator for the second order RM code of length 2^m. (Weight enumerator for second-order Reed-Muller codes, IEEE Trans. on Information Theorem, IT-16, 1970, 745-751). By the same kind of methods T. Kasami and N. Tokura have enumerated the code words with weight $< 2^{m-r+1}$ in the r-th order RM-code.

The smallest parameters for which one has not been able to find the weight enumerator are $m = 8$, $r = 3$. If $\sum_{i=0}^{256} A_i z^i$ is the weight enumerator of the 3-rd order RM code of length 256 then we know that $A_i = A_{256-i}$. From the results of Kasami and Tokura we can find A_i for $i < 64$. By (6.2.3) $A_i = 0$ if $i \not\equiv 0 \pmod 4$. This leaves 17 coefficients to be determined. Since the dual of this code is the 4-th order RM code we know B_j for $j < 32$ in (6.1.3) but this gives us only 16 independent equations. Because of the size of the A_i this problem has eluded attempts to solve it with the aid of a computer.

6.3 The Carlitz-Uchiyama bound

In 1968 D. R. Anderson observed that a result of L. Carlitz and S. Uchiyama (Bounds for exponential sums, Duke Math. J. 24 (1957), 37-41) which depended on A. Weil's well known paper on the Riemann Hypothesis in function fields (Proc. Nat. Acad. Sci. U.S.A. 34, 204-207) could be used to prove a theorem on weights in binary codes.

Consider a binary cyclic code of block length $n = 2^m - 1$. Let α be a primitive element of $GF(2^m)$. If $\underline{c}$ is a code word in this code define

$$(6.3.1) \qquad f(x) := n^{-1} \sum_{j=0}^{n-1} S_j x^j = MS(\underline{c}, x^{-1}) = n^{-1} \sum_i{}^{*} T_{\deg(\alpha^i)} (S_i x^i),$$

where

(6.3.2)
$$T_k(z) := \sum_{j=0}^{k-1} z^{2^j}$$

(see (3.4.4) and (3.4.8)). We know from (3.4.5) that if $c(x) = \sum_{i=0}^{n-1} c_i x^i$ then $c_\ell = f(\alpha^{-\ell})$. Now define

(6.3.3)
$$\check{f}(x) := n^{-1} \sum_i{}^{*} T_m(S_i x^i) = n^{-1} T_m\Big(\sum_i{}^{*} S_i x^i\Big) .$$

For this function we have the following lemma:

(6.3.4) **LEMMA:** If $c(x) = \sum_{i=0}^{n-1} c_i x^i$ is a code word in the dual of the binary t-error-correcting BCH code of block length $n = 2^m - 1$ and if $2t - 2 < 2^{m/2}$ then $c_\ell = \check{f}(\alpha^{-\ell})$ for $\ell = 0, 1, \ldots, n-1$ where α is a primitive element in $GF(2^m)$.

<u>Proof</u>: By (4.1.1) and (3.4.4) S_i differs from 0 only if i belongs to a cycle of π_2 containing one of the integers 1, 2, ..., 2t. Since, by (3.4.8), we were to choose one i from each such cycle we can make this choice in such a way that $\sum_i^* S_i x^i$ has degree $\leq 2t - 1$. To prove the lemma we must show that $f(x) = \check{f}(x)$ if $2t - 2 < 2^{m/2}$. To do this we must prove that $T_m(S_i x^i) = T_{\deg(\alpha^i)}(S_i x^i)$ for $i < 2^{\frac{1}{2}m} + 1$. Let $d_i := \deg(\alpha^i)$. By (6.3.2) and the definition of deg (α^i) it follows that $T_m(S_i x^i) = T_{d_i}(S_i x^i)$ unless m/d_i is even, say $m = 2k_i d_i$. Then, since $(2^{d_i}-1)i \equiv 0 \pmod{2^m-1}$ (because $(\alpha^i)^{2^{d_i}-1} = 1$) we have $i \geq (2^{k_i d_i} - 1)/(2^{d_i} - 1) \geq 2^{k_i d_i} + 1 = 2^{\frac{1}{2}m} + 1$. This proves the lemma.

Now the theorem of Carlitz and Uchiyama is:

(6.3.5) **THEOREM:** <u>If α is a primitive element of $GF(2^m)$ and if $g(x)$ is a polynomial of degree ν over $GF(2)$ such that for every polynomial $r(x)$ over any field $GF(2^\ell)$ the polynomial $[r(x)]^2 - r(x) - g(x)$ is not constant then</u>

$$\left| (-1)^{T_m(g(0))} + \sum_{j=1}^{2^m-1} (-1)^{T_m(g(\alpha^j))} \right| \leq (\nu-1)2^{m/2} .$$

This theorem on exponential sums over finite fields implies the following theorem:

(6.3.6) <u>THEOREM</u>: <u>The minimum distance of the dual of the binary</u> t-<u>error-correcting</u> BCH <u>code of block length</u> $n = 2^m - 1$ <u>is at least</u> $2^{m-1}-1-(t-1)2^{m/2}$.

<u>Proof</u>: Let $g(x) := \sum_i^* S_i x^i$. Let $c(x) = \sum_{i=0}^{n-1} c_i x^i \neq \underline{0}$ be a code word in the dual of the t-error-correcting BCH code. It is easily seen that the conditions of (6.3.5) are satisfied. Then from (6.3.5), (6.3.3) and (6.3.4) and the fact that n is odd we find

$$\left| 1 + \sum_{\ell=0}^{n-1} (-1)^{c_\ell} \right| \leq (2t-2)2^{m/2} .$$

This implies that the minimum weight of the code is at least $2^{m-1} - 1 - (t-1)2^{m/2}$. To apply (6.3.9) we needed the restriction $2t - 2 < 2^{m/2}$ but this can now be dropped because (6.3.6) is trivially correct if $(2t-2) \geq 2^{m/2}$.

<u>Example</u>: Consider the binary 5-error-correcting BCH code of block length 127. By inspection of the check polynomial of this code we see that the dual code is a subcode of a 7-error-correcting BCH code and therefore has minimum distance $d \geq 15$. However (6.3.6) implies $d \geq 18$.

Knowledge of the type provided by (6.3.6) can be very useful in weight enumeration e.g. when one wishes to apply the MacWilliams relations (6.1.3).

References

[1] E. F. Assmus and H. F. Mattson, On tactical configurations and error-correcting codes, Journal of Comb. Theory 3 (1967), 243-257.

[2] E. F. Assmus, H. F. Mattson, R. Turyn, Cyclic codes, Report AFCRL-65-332 of the Applied Research Laboratory of Sylvania Electronic Systems (1965).

[3] E. F. Assmus, H. F. Mattson, R. Turyn, Cyclic codes, Report AFCRL-66-348 of the Applied Research Laboratory of Sylvania Electronic Systems (1966).

[4] E. F. Assmus, H. F. Mattson, R. Turyn, Research to develop the algebraic theory of codes, Report AFCRL-67-0365 of the Applied Research Laboratory of Sylvania Electronic Systems (1967).

[5] E. R. Berlekamp, Algebraic coding theory, McGraw Hill, New York (1968).

[6] R. G. Gallager, Information theory and reliable communication, Wiley, New York (1968).

[7] H. J. L. Kamps and J. H. van Lint, The football pool problem for 5 matches, Journal of Comb. Theory 3 (1967), 315-325.

[8] A. I. Khinchin, Mathematical foundations of information theory, Dover, New York (1957).

[9] J. H. van Lint, 1967-1969 Report of the discrete mathematics group, Report 69-WSK-04 of the Technological University, Eindhoven, Netherlands (1969).

[10] J. H. van Lint, On the nonexistence of perfect 2- and 3- Hamming-error-correcting codes over GF(q), Information and Control 16 (1970), 396-401.

[11] J. H. van Lint, Nonexistence theorems for perfect error-correcting codes, Computers in algebra and number theory, SIAM-AMS Vol. 3 (to appear).

[12] J. H. van Lint, Nonexistence of perfect 5-6-and 7-Hamming-error-correcting codes over GF(q), Report 70-WSK-06 of the Technological University Eindhoven, Netherlands (1970).

[13] S. P. Lloyd, Binary block coding, Bell System Tech. J. 36 (1957), 517-535.

[14] H. F. Mattson and G. Solomon, A new treatment of Bose-Chaudhuri codes, J. Soc. Indust. Appl. Math. 9 (1961), 654-669.

[15] W. W. Peterson, Error-correcting codes, M.I.T. Press, Cambridge (1961).

[16] I. S. Reed, A class of multiple-error-correcting codes and the decoding scheme, IEEE Trans. Inform. Theory, IT-4 (1954), 38-49.

[17] D. Slepian, Coding theory, Nuovo Cimento 13 (1959), 373-388.

[18] A. D. Wyner, On coding and information theory, SIAM Review 11 (1969), 317-346.

[19] A. Tietäväinen, On the non-existence of perfect codes over finite fields, SIAM j. Appl. Math. 24 (1973), 88-96.

[20] W.A. Zinoviev and W.K. Leontiev, A theorem on the nonexistence of perfect codes over finite fields (Russian), to appear.

[21] H.W. Lenstra jr., Two theorems on perfect codes, Discrete Math. 3 (1972), 125-132.

[22] J.H. van Lint, A survey of perfect codes, to appear in Rocky Mountain J. Math.

INDEX

Lecture Notes in Mathematics

Comprehensive leaflet on request

Vol. 146: A. B. Altman and S. Kleiman, Introduction to Grothendieck Duality Theory. II, 192 pages. 1970. DM 18,–

Vol. 147: D. E. Dobbs, Cech Cohomological Dimensions for Commutative Rings. VI, 176 pages. 1970. DM 16,–

Vol. 148: R. Azencott, Espaces de Poisson des Groupes Localement Compacts. IX, 141 pages. 1970. DM 16,–

Vol. 149: R. G. Swan and E. G. Evans, K-Theory of Finite Groups and Orders. IV, 237 pages. 1970. DM 20,–

Vol. 150: Heyer, Dualität lokalkompakter Gruppen. XIII, 372 Seiten. 1970. DM 20,–

Vol. 151: M. Demazure et A. Grothendieck, Schémas en Groupes I. (SGA 3). XV, 562 pages. 1970. DM 24,–

Vol. 152: M. Demazure et A. Grothendieck, Schémas en Groupes II. (SGA 3). IX, 654 pages. 1970. DM 24,–

Vol. 153: M. Demazure et A. Grothendieck, Schémas en Groupes III. (SGA 3). VIII, 529 pages. 1970. DM 24,–

Vol. 154: A. Lascoux et M. Berger, Variétés Kähleriennes Compactes. VII, 83 pages. 1970. DM 16,–

Vol. 155: Several Complex Variables I, Maryland 1970. Edited by J. Horváth. IV, 214 pages. 1970. DM 18,–

Vol. 156: R. Hartshorne, Ample Subvarieties of Algebraic Varieties. XIV, 256 pages. 1970. DM 20,–

Vol. 157: T. tom Dieck, K. H. Kamps und D. Puppe, Homotopietheorie. VI, 265 Seiten. 1970. DM 20,–

Vol. 158: T. G. Ostrom, Finite Translation Planes. IV. 112 pages. 1970. DM 16,–

Vol. 159: R. Ansorge und R. Hass. Konvergenz von Differenzenverfahren für lineare und nichtlineare Anfangswertaufgaben. VIII, 145 Seiten. 1970. DM 16,–

Vol. 160: L. Sucheston, Constributions to Ergodic Theory and Probability. VII, 277 pages. 1970. DM 20,–

Vol. 161: J. Stasheff, H-Spaces from a Homotopy Point of View. VI, 95 pages. 1970. DM 16,–

Vol. 162: Harish-Chandra and van Dijk, Harmonic Analysis on Reductive p-adic Groups. IV, 125 pages. 1970. DM 16,–

Vol. 163: P. Deligne, Equations Différentielles à Points Singuliers Reguliers. III, 133 pages. 1970. DM 16,–

Vol. 164: J. P. Ferrier, Seminaire sur les Algebres Complètes. II, 69 pages. 1970. DM 16,–

Vol. 165: J. M. Cohen, Stable Homotopy. V, 194 pages. 1970. DM 16,–

Vol. 166: A. J. Silberger, PGL_2 over the p-adics: its Representations, Spherical Functions, and Fourier Analysis. VII, 202 pages. 1970. DM 18,–

Vol. 167: Lavrentiev, Romanov and Vasiliev, Multidimensional Inverse Problems for Differential Equations. V, 59 pages. 1970. DM 16,–

Vol. 168: F. P. Peterson, The Steenrod Algebra and its Applications: A conference to Celebrate N. E. Steenrod's Sixtieth Birthday. VII, 317 pages. 1970. DM 22,–

Vol. 169: M. Raynaud, Anneaux Locaux Henséliens. V, 129 pages. 1970. DM 16,–

Vol. 170: Lectures in Modern Analysis and Applications III. Edited by C. T. Taam. VI, 213 pages. 1970. DM 18,–

Vol. 171: Set-Valued Mappings, Selections and Topological Properties of 2^X. Edited by W. M. Fleischman. X, 110 pages. 1970. DM 16,–

Vol. 172: Y.-T. Siu and G. Trautmann, Gap-Sheaves and Extension of Coherent Analytic Subsheaves. V, 172 pages. 1971. DM 16,–

Vol. 173: J. N. Mordeson and B. Vinograde, Structure of Arbitrary Purely Inseparable Extension Fields. IV, 138 pages. 1970. DM 16,–

Vol. 174: B. Iversen, Linear Determinants with Applications to the Picard Scheme of a Family of Algebraic Curves. VI, 69 pages. 1970. DM 16,–

Vol. 175: M. Brelot, On Topologies and Boundaries in Potential Theory. VI, 176 pages. 1971. DM 18,–

Vol. 176: H. Popp, Fundamentalgruppen algebraischer Mannigfaltigkeiten. IV, 154 Seiten. 1970. DM 16,–

Vol. 177: J. Lambek, Torsion Theories, Additive Semantics and Rings of Quotients. VI, 94 pages. 1971. DM 16,–

Vol. 178: Th. Bröcker und T. tom Dieck, Kobordismentheorie. XVI, 191 Seiten. 1970. DM 18,–

Vol. 179: Seminaire Bourbaki – vol. 1968/69. Exposés 347-363. IV. 295 pages. 1971. DM 22,–

Vol. 180: Séminaire Bourbaki – vol. 1969/70. Exposés 364-381. IV, 310 pages. 1971. DM 22,–

Vol. 181: F. DeMeyer and E. Ingraham, Separable Algebras over Commutative Rings. V, 157 pages. 1971. DM 16.–

Vol. 182: L. D. Baumert. Cyclic Difference Sets. VI, 166 pages. 1971. DM 16,–

Vol. 183: Analytic Theory of Differential Equations. Edited by P. F. Hsieh and A. W. J. Stoddart. VI, 225 pages. 1971. DM 20,–

Vol. 184: Symposium on Several Complex Variables, Park City, Utah, 1970. Edited by R. M. Brooks. V, 234 pages. 1971. DM 20,–

Vol. 185: Several Complex Variables II, Maryland 1970. Edited by J. Horváth. III, 287 pages. 1971. DM 24,–

Vol. 186: Recent Trends in Graph Theory. Edited by M. Capobianco/ J. B. Frechen/M. Krolik. VI, 219 pages. 1971. DM 18.–

Vol. 187: H. S. Shapiro, Topics in Approximation Theory. VIII, 275 pages. 1971. DM 22,–

Vol. 188: Symposium on Semantics of Algorithmic Languages. Edited by E. Engeler. VI, 372 pages. 1971. DM 26,–

Vol. 189: A. Weil, Dirichlet Series and Automorphic Forms. V, 164 pages. 1971. DM 16,–

Vol. 190: Martingales. A Report on a Meeting at Oberwolfach, May 17-23, 1970. Edited by H. Dinges. V, 75 pages. 1971. DM 16,–

Vol. 191: Séminaire de Probabilités V. Edited by P. A. Meyer. IV, 372 pages. 1971. DM 26,–

Vol. 192: Proceedings of Liverpool Singularities – Symposium I. Edited by C. T. C. Wall. V, 319 pages. 1971. DM 24,–

Vol. 193: Symposium on the Theory of Numerical Analysis. Edited by J. Ll. Morris. VI, 152 pages. 1971. DM 16,–

Vol. 194: M. Berger, P. Gauduchon et E. Mazet. Le Spectre d'une Variété Riemannienne. VII, 251 pages. 1971. DM 22,–

Vol. 195: Reports of the Midwest Category Seminar V. Edited by J.W. Gray and S. Mac Lane. III, 255 pages. 1971. DM 22,–

Vol. 196: H-spaces – Neuchâtel (Suisse)- Août 1970. Edited by F. Sigrist, V, 156 pages. 1971. DM 16,–

Vol. 197: Manifolds – Amsterdam 1970. Edited by N. H. Kuiper. V, 231 pages. 1971. DM 20,–

Vol. 198: M. Hervé, Analytic and Plurisubharmonic Functions in Finite and Infinite Dimensional Spaces. VI, 90 pages. 1971. DM 16.–

Vol. 199: Ch. J. Mozzochi, On the Pointwise Convergence of Fourier Series. VII, 87 pages. 1971. DM 16,–

Vol. 200: U. Neri, Singular Integrals. VII, 272 pages. 1971. DM 22,–

Vol. 201: J. H. van Lint, Coding Theory. VII, 136 pages. 1971. DM 16,–

Vol. 202: J. Benedetto, Harmonic Analysis on Totally Disconnected Sets. VIII, 261 pages. 1971. DM 22,–

Vol. 203: D. Knutson, Algebraic Spaces. VI, 261 pages. 1971. DM 22,–

Vol. 204: A. Zygmund, Intégrales Singulières. IV, 53 pages. 1971. DM 16,–

Vol. 205: Séminaire Pierre Lelong (Analyse) Année 1970. VI, 243 pages. 1971. DM 20,–

Vol. 206: Symposium on Differential Equations and Dynamical Systems. Edited by D. Chillingworth. XI, 173 pages. 1971. DM 16,–

Vol. 207: L. Bernstein, The Jacobi-Perron Algorithm – Its Theory and Application. IV, 161 pages. 1971. DM 16,–

Vol. 208: A. Grothendieck and J. P. Murre, The Tame Fundamental Group of a Formal Neighbourhood of a Divisor with Normal Crossings on a Scheme. VIII, 133 pages. 1971. DM 16,–

Vol. 209: Proceedings of Liverpool Singularities Symposium II. Edited by C. T. C. Wall. V, 280 pages. 1971. DM 22,–

Vol. 210: M. Eichler, Projective Varieties and Modular Forms. III, 118 pages. 1971. DM 16,–

Vol. 211: Théorie des Matroïdes. Edité par C. P. Bruter. III, 108 pages. 1971. DM 16,–

Please turn over

Vol. 278: H. Jacquet, Automorphic Forms on GL(2). Part II. XIII, 142 pages. 1972. DM 16,–

Vol. 279: R. Bott, S. Gitler and I. M. James, Lectures on Algebraic and Differential Topology. V, 174 pages. 1972. DM 18,–

Vol. 280: Conference on the Theory of Ordinary and Partial Differential Equations. Edited by W. N. Everitt and B. D. Sleeman. XV, 367 pages. 1972. DM 26,–

Vol. 281: Coherence in Categories. Edited by S. Mac Lane. VII, 235 pages. 1972. DM 20,–

Vol. 282: W. Klingenberg und P. Flaschel, Riemannsche Hilbertmannigfaltigkeiten. Periodische Geodätische. VII, 211 Seiten. 1972. DM 20,–

Vol. 283: L. Illusie, Complexe Cotangent et Déformations II. VII, 304 pages. 1972. DM 24,–

Vol. 284: P. A. Meyer, Martingales and Stochastic Integrals I. VI, 89 pages. 1972. DM 16,–

Vol. 285: P. de la Harpe, Classical Banach-Lie Algebras and Banach-Lie Groups of Operators in Hilbert Space. III, 160 pages. 1972. DM 16,–

Vol. 286: S. Murakami, On Automorphisms of Siegel Domains. V, 95 pages. 1972. DM 16,–

Vol. 287: Hyperfunctions and Pseudo-Differential Equations. Edited by H. Komatsu. VII, 529 pages. 1973. DM 36,–

Vol. 288: Groupes de Monodromie en Géométrie Algébrique. (SGA 7 I). Dirigé par A. Grothendieck. IX, 523 pages. 1972. DM 50,–

Vol. 289: B. Fuglede, Finely Harmonic Functions. III, 188. 1972. DM 18,–

Vol. 290: D. B. Zagier, Equivariant Pontrjagin Classes and Applications to Orbit Spaces. IX, 130 pages. 1972. DM 16,–

Vol. 291: P. Orlik, Seifert Manifolds. VIII, 155 pages. 1972. DM 16,–

Vol. 292: W. D. Wallis, A. P. Street and J. S. Wallis, Combinatorics: Room Squares, Sum-Free Sets, Hadamard Matrices. V, 508 pages. 1972. DM 50,–

Vol. 293: R. A. DeVore, The Approximation of Continuous Functions by Positive Linear Operators. VIII, 289 pages. 1972. DM 24,–.

Vol. 294: Stability of Stochastic Dynamical Systems. Edited by R. F. Curtain. IX, 332 pages. 1972. DM 26,–

Vol. 295: C. Dellacherie, Ensembles Analytiques, Capacités, Mesures de Hausdorff. XII, 123 pages. 1972. DM 16,–

Vol. 296: Probability and Information Theory II. Edited by M. Behara, K. Krickeberg and J. Wolfowitz. V, 223 pages. 1973. DM 20,–

Vol. 297: J. Garnett, Analytic Capacity and Measure. IV, 138 pages. 1972. DM 16,–

Vol. 298: Proceedings of the Second Conference on Compact Transformation Groups. Part 1. XIII, 453 pages. 1972. DM 32,–

Vol. 299: Proceedings of the Second Conference on Compact Transformation Groups. Part 2. XIV, 327 pages. 1972. DM 26,–

Vol. 300: P. Eymard, Moyennes Invariantes et Représentations Unitaires. II. 113 pages. 1972. DM 16,–

Vol. 301: F. Pittnauer, Vorlesungen über asymptotische Reihen. VI, 186 Seiten. 1972. DM 18,–

Vol. 302: M. Demazure, Lectures on p-Divisible Groups. V, 98 pages. 1972. DM 16,–

Vol. 303: Graph Theory and Applications. Edited by Y. Alavi, D. R. Lick and A. T. White. IX, 329 pages. 1972. DM 26,–

Vol. 304: A. K. Bousfield and D. M. Kan, Homotopy Limits, Completions and Localizations. V, 348 pages. 1972. DM 26,–

Vol. 305: Théorie des Topos et Cohomologie Etale des Schémas. Tome 3. (SGA 4). Dirigé par M. Artin, A. Grothendieck et J. L. Verdier. VI, 640 pages. 1973. DM 50,–

Vol. 306: H. Luckhardt, Extensional Gödel Functional Interpretation. VI, 161 pages. 1973. DM 18,–

Vol. 307: J. L. Bretagnolle, S. D. Chatterji et P.-A. Meyer, Ecole d'été de Probabilités: Processus Stochastiques. VI, 198 pages. 1973. DM 20,–

Vol. 308: D. Knutson, λ-Rings and the Representation Theory of the Symmetric Group. IV, 203 pages. 1973. DM 20,–

Vol. 309: D. H. Sattinger, Topics in Stability and Bifurcation Theory. VI, 190 pages. 1973. DM 18,–

Vol. 310: B. Iversen, Generic Local Structure of the Morphisms in Commutative Algebra. IV, 108 pages. 1973. DM 16,–

Vol. 311: Conference on Commutative Algebra. Edited by J. W. Brewer and E. A. Rutter. VII, 251 pages. 1973. DM 22,–

Vol. 312: Symposium on Ordinary Differential Equations. Edited by W. A. Harris, Jr. and Y. Sibuya. VIII, 204 pages. 1973. DM 22,–

Vol. 313: K. Jörgens and J. Weidmann, Spectral Properties of Hamiltonian Operators. III, 140 pages. 1973. DM 16,–

Vol. 314: M. Deuring, Lectures on the Theory of Algebraic Functions of One Variable. VI, 151 pages. 1973. DM 16,–

Vol. 315: K. Bichteler, Integration Theory (with Special Attention to Vector Measures). VI, 357 pages. 1973. DM 26,–

Vol. 316: Symposium on Non-Well-Posed Problems and Logarithmic Convexity. Edited by R. J. Knops. V, 176 pages. 1973. DM 18,–

Vol. 317: Séminaire Bourbaki – vol. 1971/72. Exposés 400–417. IV, 361 pages. 1973. DM 26,–

Vol. 318: Recent Advances in Topological Dynamics. Edited by A. Beck, VIII, 285 pages. 1973. DM 24,–

Vol. 319: Conference on Group Theory. Edited by R. W. Gatterdam and K. W. Weston. V, 188 pages. 1973. DM 18,–

Vol. 320: Modular Functions of One Variable I. Edited by W. Kuyk. V, 195 pages. 1973. DM 18,–

Vol.: 321: Séminaire de Probabilités VII. Edité par P. A. Meyer. VI, 322 pages. 1973. DM 26,–

Vol. 322: Nonlinear Problems in the Physical Sciences and Biology. Edited by I. Stakgold. D. D. Joseph and D. H. Sattinger, VIII, 357 pages. 1973. DM 26,–

Vol. 323: J. L. Lions, Perturbations Singulières dans les Problèmes aux Limites et en Contrôle Optimal. XII, 645 pages. 1973. DM 42,–

Vol. 324: K. Kreith, Oscillation Theory. VI, 109 pages. 1973. DM 16,–

Vol. 325: Ch.-Ch. Chou, La Transformation de Fourier Complexe et L'Equation de Convolution. IX, 137 pages. 1973. DM 16,–

Vol. 326: A. Robert, Elliptic Curves. VIII, 264 pages. 1973. DM 22,–

Vol. 327: E. Matlis, 1-Dimensional Cohen-Macaulay Rings. XII, 157 pages. 1973. DM 18,–

Vol. 328: J. R. Büchi and D. Siefkes, The Monadic Second Order Theory of All Countable Ordinals. VI, 217 pages. 1973. DM 20,–

Vol. 329: W. Trebels, Multipliers for (C, α)-Bounded Fourier Expansions in Banach Spaces and Approximation Theory. VII, 103 pages. 1973. DM 16,–

Vol. 330: Proceedings of the Second Japan-USSR Symposium on Probability Theory. Edited by G. Maruyama and Yu. V. Prokhorov. VI, 550 pages. 1973. DM 36,–

Vol. 331: Summer School on Topological Vector Spaces. Edited by L. Waelbroeck. VI, 226 pages. 1973. DM 20,–

Vol. 332: Séminaire Pierre Lelong (Analyse) Année 1971–1972. V, 131 pages. 1973. DM 16,–

Vol. 333: Numerische, insbesondere approximationstheoretische Behandlung von Funktionalgleichungen. Herausgegeben von R. Ansorge und W. Törnig. VI, 296 Seiten. 1973. DM 24,–

Vol. 334: F. Schweiger, The Metrical Theory of Jacobi-Perron Algorithm. V, 111 pages. 1973. DM 16,–

Vol. 335: H. Huck, R. Roitzsch, U. Simon, W. Vortisch, R. Walden, B. Wegner und W. Wendland, Beweismethoden der Differentialgeometrie im Großen. IX, 159 Seiten. 1973. DM 18,–

Vol. 336: L'Analyse Harmonique dans le Domaine Complexe. Edité par E. J. Akutowicz. VIII, 169 pages. 1973. DM 18,–